Computer Simulation
of
Electronic Circuits

Computer Simulation
of
Electronic Circuits

R. RAGHURAM

Department of Electrical Engineering
Indian Institute of Technology, Kanpur
India

JOHN WILEY & SONS

New York   Chichester   Brisbane   Toronto   Singapore

First Published in 1989 by
WILEY EASTERN LIMITED
4835/24 Ansari Road, Daryaganj
New Delhi 110 002, India

Distributors :

Australia and New Zealand:
JACARANDA WILEY LTD.
GPO Box 859, Brisbane, Queensland 4001, Australia

Canada:
JOHN WILEY & SONS CANADA LIMITED
22 Worcester Road, Rexdale, Ontario, Canada

Europe and Africa:
JOHN WILEY & SONS LIMITED
Baffins lane, Chichester, West Sussex, England

South East Asia:
JOHN WILEY & SONS (PTE) LTD.
05-04, Block B, Union Industrial Building
37 Jalan Pemimpin, Singapore 2057

Africa and South Asia:
WILEY EASTERN LIMITED
4835/24 Ansari Road, Daryaganj
New Delhi 110 002, India

North and South America and rest of the World:
JOHN WILEY & SONS, INC.
605 Third Avenue, New York, NY 10158, USA

Library of Congress of Cataloging-in-Publication Data
Raghuram, R.
    Computer simulation of electronic circuits/R. Raghuram.
      p. cm.
    Includes index
    1. Electronic circuits—Computer simulation.   I.   Title
       TK7867.R27   1989   88-29772
       621.3815˙3'0724—dc19   CIP

ISBN 0-470-21331-0 John Wiley & Sons, Inc.
ISBN 81-224-0111-2 Wiley Eastern Limited

Printed in India at Rajkamal Electric Press, Delhi.

Preface

Students of Electrical Engineering are often exposed at an early stage, nowadays, to simulation of circuits on a computer. Sophisticated software packages are available with universities and semiconductor houses. The aim of this book is to help students or engineers use these packages more effectively to design circuits. The underlying theory is described in adequate detail that students can develop their own packages or modify existing ones if necessary. Design courses on integrated circuits rely heavily on the availability of such packages.

Most books on computer aided design stress only one of the following-network analysis or device modelling or digital (logic) simulation. This book aims at providing an integrated approach. A network analysis program is useless without good models for common devices like diodes, BJTs and MOSFETs. In packages where good models are already implemented, a proper knowledge of the various parameters is necessary. Large digital circuits need logic simulation at least as a preliminary step. Timing simulators are increasingly being used to bridge the gap between logic and circuit simulation. Here again, a knowledge of network analysis, device modelling and logic simulation is imperative. Hence the need for the integrated approach.

The background required is that of a final year UG student in Electrical Engineering. Mathematics has been kept to a minimum and an intuitive rather than a rigorous approach adopted since a wide range of topics is covered. A number of solved and unsolved examples, programs and program segments have been included wherever found necessary. It is presumed that the student has access to a computer system. Selected portions of the book may also be used to supplement other reading material in courses on analog and digital circuits, semiconductor device theory and network theory.

The book can be completed in one semester. Depending on the nature of the course one can highlight either the network or modelling or digital circuit simulation part of the course. Problems have been listed at the end of each chapter and some of them are programming assignments.

This book would not have been possible but for the considerable feedback from students over the past few years. I was fortunate in having a reasonably large class with some excellent students who streamlined my thinking and presentation. The various computer programs described here are from the theses of S Banerjee, R S Raghunandan, R R Prasad and F Rozario. It has been a pleasure to work with such wonderful students.

My sincere thanks to Mr J S Rawat for typing, Mr D K Mishra for drafting the manuscript and Mr Triveni Tiwari and Mr Ganga Ram for running off the copies. Mr T Sadagopan has played the role of a production manager overseeing everything from the quality of the paper to speedy dispatch of the manuscript. The mistakes remaining in the book are despite their patient efforts and the author readily accepts responsibility for them. Of course, the financial assistance provided by the QIP program was the prime mover behind the whole effort.

R. RAGHURAM

List of Symbols

$[A]$	Matrix in solving a system of linear equations $[A][x] = [B]$
$[B]$	Known vector in solving a system of linear equations $[A][x] = [B]$
B	Base terminal of a BJT; also substrate terminal of MOSFET
C	Collector terminal of BJT
C_T	Transition or depletion region capacitance of a pn junction
C_{XY}	Capacitance between terminals X and Y
C_D	Diffusion capacitance of a pn junction
D_P	Diffusion Coefficient for holes
D_n	Diffusion Coefficient for electrons
D	Drain Terminal of a MOSFET
E_F	Fermi Level
E	Emitter terminal of a BJT
f	Frequency
g_m	Transconductance of a BJT
G	Gate terminal of a MOSFET; also any conductance
I	DC current
$i(t)$	Current which is time-varying
I	Sinusoidal current represented by a phasor
J, j	Current density; j is also $(-1)^{1/2}$
k	Boltzmann constant
$[L]$	Upper triangular matrix after LU Decomposition
l	Length of channel in a MOSFET
m	Grading coefficient for pn junction
n	Semiconductor doped such that mobile carriers are mainly electrons
p	Semiconductor doped such that mobile carriers are mainly holes
q, Q	Charge; q is also charge of an electron
r	Resistance
$[S]$	Sensitivity matrix
T	Temperature in degrees Kelvin
$[U]$	Upper triangular matrix after LU Decomposition
$[u]$	Vector of known time varying inputs
V	DC voltage
V	Sinusoidal voltage represented by a phasor
V_T	Same as kT/q which is 26 mV at 300 degrees Kelvin
$v(t)$	Voltage which is time-varying
W	Width of the channel in a MOSFET

$[x]$	Vector of unknowns in solving simultaneous equations—linear, non–linear or differential
$[Y]$	Node admittance matrix
$[Z]$	Loop impedance matrix
α_F	Parameter of Ebers-Moll model of a BJT roughly equal to I_C/I_E in the normal active mode
α_R	Parameter of Ebers-Moll model of a BJT roughly equal to I_E/I_C in the inverse active mode
β_F	Parameter of Ebers-Moll model of a BJT roughly equal to I_C/I_B in the normal active mode
β_R	Parameter of Ebers-Moll model of a BJT roughly equal to I_E/I_B in the inverse active mode
γ	Bulk threshold parameter for a MOSFET
Δ	Incremental change in any quantity
ϵ	Dielectric constant
η	Emission constant for a *pn* junction
λ	Flux linkage; also channel length modulation parameter for a MOSFET
μ_n	Mobility for holes
μ_P	Mobility for electrons
ρ	Resistivity
σ	Conductivity
τ_P	Lifetime for holes
τ_n	Lifetime for electrons
τ_t	Transit time
ϕ_o	Zero bias contact potential for *pn* junction
ϕ_F	Fermi potential
ω	Angular frequency

Contents

1

Introduction and Overview

1.1 Role of Simulation in Integrated Circuit (IC) Design

Most of us realise sometime or the other how tiresome it is to solve circuits, having as few as 2 or 3 transistors. For example, the common TTL inverter, Fig. 1.1(a) which has only 4 transistors, is difficult to analyse even for DC inputs. The step response, taking all the non-linear capacitances of each transistor into account, is impossible to obtain by hand calculations. Even in the case of an NMOS inverter Fig. 1.1(b), which has only two MOSFETS, it is not possible. Today we have VLSIs with hundreds of thousands of transistors, which require a computer to simulate circuit behaviour. Breadboarding is not a viable option, since a device behaves differently in a chip. Further circuit complexity has really gone far beyond the bread-boarding level.

Figure 1.2 shows the steps in the design of a typical MOS digital circuit, beginning with the specifications. In a microprocessor chip, this would mean the number of registers, instruction set, interrupt facilities, etc. The functional or behavioural level design and perhaps simulation follows [1]. Register transfer languages (RTL) like AHPL [2] (see also Section 8.9) may be used for simulation at the next lower level, followed by a gate level design. A gate level logic simulation is used to verify this. More critical parts of the circuit are simulated by a circuit simulator [3]. After each simulation there is normally some feedback and change in the design. The circuit is then actually laid out using graphics facilities, pattern generator tapes made, masks 'cut' and finally chips fabricated, packed and tested. Testing activity is related to simulation, as described in Chapter 8. The designer is expected to generate the test patterns and chips are usually designed keeping testability in mind. Most of the steps in Fig. 1.2 are automated. One even talks of silicon compilers which generate masks automatically given the functional specifications.

This work is confined to the simulation of the circuit at various levels—circuit, logic, etc. Simulation can also be used to analyse and design circuits using discrete devices/chips and most of the techniques are also valid for such circuits. Models for common semiconductor devices and integrated circuits are a part of simulation and are, therefore, very much part of this book.

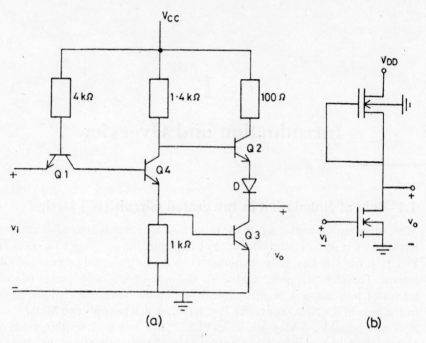

Fig. 1.1   (a) TTL inverter (b) NMOS inverter with depletion mode load transistor.

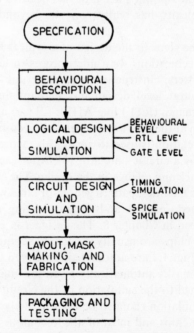

Fig. 1.2   Various steps in manufacture of
a MOS digital circuit.

1.2 Various Kinds of Elements

The network elements one comes across can be variously categorised—according to linearity, type of non-linearity, number of terminals, active or passive, dissipative or storage, time-varying or constant, etc. In this book, multi-terminal elements are always reduced to two terminal elements. For example, the three terminal transistor is represented by an equivalent circuit or model (like the Ebers-Moll model) containing only two terminal elements. The basic elements one deals with are resistors, inductors, capacitors and four kinds of controlled sources. The four kinds of controlled sources are:

1. Voltage controlled voltage source (VCVS)
2. Current controlled voltage source (CCVS)
3. Voltage controlled current source (VCCS)
4. Current controlled current source (CCCS)

A network or circuit (both terms will be used interchangeably) also has exciting current and voltage sources. These are functions of time, in general, and called independent sources to distinguish them from the dependent or controlled sources listed above. For purposes of analysis, controlled sources should be considered as network elements and not as excitations.

All the network elements listed above can be linear or non-linear. The definition of linearity of a network element is similar to that of a system. A linear element has to satisfy the properties of homogeneity and additivity.

Homogeneity: If the response to an input $I(t)$ is $R(t)$, then the response to an input $KI(t)$ is $KR(t)$. This is also known as the scaling principle.

Additivity: If the response to input $I_1(t)$ is $R_1(t)$ and that due to $I_2(t)$ is $R_2(t)$ then the response to the input $(I_1(t) + I_2(t))$ is $(R_1(t) + R_2(t))$.

In network elements, we can consider currents as inputs and voltages as responses or vice versa. Figure 1.3 shows the i–v characteristics of 3 resistors. The resistor represented by Fig. 1.3(a) satisfies both homogeneity and additivity and is therefore a linear resistor. The resistor represented by Fig. 1.3(b) satisfies only homogeneity while the one in Fig. 1.3(c) satisfies neither homogeneity nor additivity. Both of these are therefore non-linear.

If a resistor is non-linear, it may be current controlled or voltage controlled. Figure 1.4(a) shows a current controlled non-linear resistor and Fig. 1.4(b) a voltage controlled non-linear resistor. Figure 1.4(c) shows an i–v characteristics which may be considered current or voltage controlled. In a current controlled non-linear element, the specifying of current uniquely specifies the voltage while the opposite is not necessarily true. A voltage controlled non-linear element, exhibits a dual behaviour. A thyristor is an example of a current controlled non-linear element, while a tunnel diode has a voltage controlled characteristic. The common p–n junction diode can be considered current or voltage controlled depending on convenience. As explained below, capacitors and inductors can also be non-linear. Non-linear capacitances in most semiconductor devices are best considered to be voltage controlled. Non-linear inductors are usually current controlled.

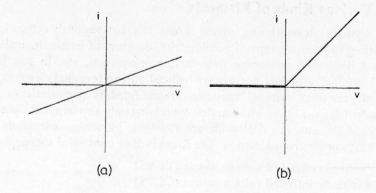

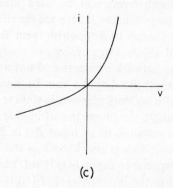

Fig. 1.3 (a) $i–v$ characteristics of a linear resistor (b) Non-linear resistor which satisfies homogeneity but not additivity, (c) Non-linear resistor which satisfies neither homogeneity nor additivity.

A capacitor's capacitance is defined by

$$C = \frac{Q}{v}$$

where Q is the charge and v the voltage across the capacitor. A linear capacitor has Q directly proportional to v. The $i–v$ general relations for any general capacitor are

$$i(t) = \frac{dQ}{dt} = \frac{d(Cv)}{dt} = \left(C + v\frac{dC}{dv}\right)\frac{dv}{dt} \tag{1.1}$$

$$v = \frac{Q}{C} = \frac{1}{C}\int_{-\infty}^{t} i(\tau)\,d\tau \tag{1.2}$$

Even for a non-linear capacitor, one would like to write

$$i(t) = C_{\text{eff}}\frac{dv}{dt} \cdot \text{From Eq. (1.1), } C_{\text{eff}}(v) = \left(C + v\frac{dC}{dv}\right) = \frac{dQ}{dv}.$$

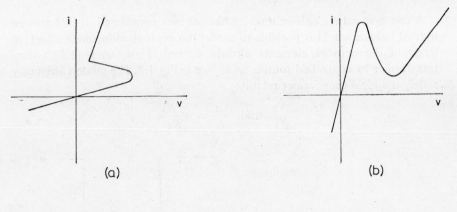

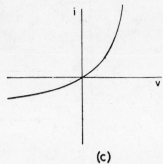

(c)

Fig. 1.4 (a) Current controlled non-linear resistor, (b) Voltage controlled non-linear resistor, (c) Non-linear resistor which can be considered as voltage or current controlled.

Similarly for a inductor, the inductance L is defined by

$$L = \frac{\lambda}{i}$$

where λ is the total number of flux linkages. For a n turn coil having flux ϕ through it, λ is equal to $n\phi$.

$$v(t) = \frac{d\lambda}{dt} = \frac{d}{dt}(Li) = \left(L + i\frac{dL}{di}\right)\frac{ai}{dt} \tag{1.3}$$

Effective inductance as seen in Eq. (1.3) is

$$L_{\text{eff}}(i) = L + i\frac{dL}{di}$$

As an example, consider the depletion or transition region capacitance of a p-n junction diode. It has

$$Q = K(\phi_0 - v)^{1/2} \tag{1.4}$$

where K is a function of the p and n region impurity concentration levels and ϕ_0 is the zero bias contact potential. From Eq. (1.4) we have

$$C_T = \frac{dQ}{dv} = \frac{K}{2(\phi_0 - v)^{1/2}} \tag{1.5}$$

This capacitance is usually important only for reverse bias, i.e. v negative.

A common network element which has not been considered so far has been mutual inductance. It is possible to model the mutual inductance effect in terms of the network elements already defined. First, we model an ideal transformer by controlled sources as shown in Fig. 1.5. The models obviously satisfy the ideal transformer relations.

$$\frac{v_2}{v_1} = \frac{N_2}{N_1} \quad \text{and} \quad \frac{i_2}{i_1} = \frac{-N_1}{N_2}$$

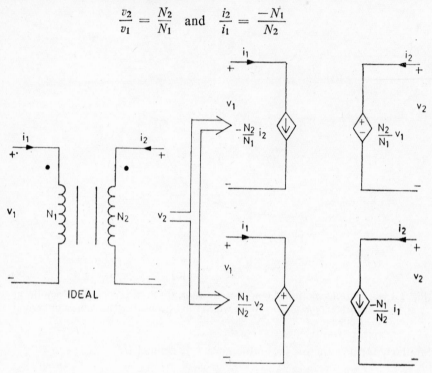

Fig. 1.5   Models for an ideal transformer.

A pair of mutually coupled coils can be modelled in terms of inductors and an ideal transformer as shown in Fig. 1.6. In Fig. 1.6(a) the ideal transformer is needed only for isolation. The models in Fig. 1.6(b) and Fig. 1.6(c) pertain to the primary and secondary coils, respectively. These are more convenient as none of the inductors can become negative. Further the series inductors $(L_1 - M')$, $(L_2' - M')$, $(L_1'' - M'')$ and $(L_2 - M'')$ can clearly be identified as leakage reactances and the parallel inductors M' and M'' as magnetising reactances.

Network elements and networks can be classified as active and passive. Practical resistors, inductors and capacitors are passive. They can dissipate or store energy, but can never supply more energy than was originally absorbed or stored. The term 'practical' has been used because an idealised negative resistance with the i–v characteristics in the second or fourth quardrants can be an active element. Independent and dependent sources (both voltage and current) are active elements capable of supplying energy.

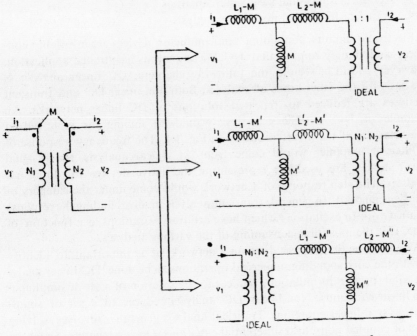

Fig. 1.6 Models for a pair of coupled coils in terms of an ideal transformer.

The original source of energy may be a chemical (battery), mechanical (generator) or an alternative electrical source (like DC supply for a dependent source amplifying an AC signal). A network may also be active or passive. The following relation may be used to check whether a network element or network is passive at a port or pair of terminals. Let $E(t_0)$ be the energy input to the port from some external source(s) upto time t_0. Let $v(t)$ and $i(t)$ represent the voltage and current at the input to the port, polarities being similar to those in Fig. 1.5. Then the network connected between the terminals of the port is passive if and only if

$$E(t) = E(t_0) + \int_{t_0}^{t} v(\tau)\, i(\tau)\, d\tau \geqslant 0$$

for any voltage, its corresponding current and for all $t > t_0$. In other words, energy delivered to the element or port must not be negative.

Network elements can be time varying. However, time varying elements, which are not necessarily non-linear, will not be considered in this book.

1.3 Different Kinds of Analyses

On a given circuit, various kinds of analyses can be done. Most packages for circuit analysis (like SPICE [3]) mostly contain the following:

1. DC linear analysis
2. DC non-linear analysis

3. AC linear sinusoidal steady state analysis
4. Transient linear and non-linear analysis

Electronic circuits containing semiconductor devices are invariably non-linear and it may appear that DC linear analysis has limited application. However, as will be seen in the following chapters, DC linear analysis is the basis for the other kinds of analyses. Both non-linear DC and transient analyses are reduced to repeated solutions of DC linear networks. An example of DC non-linear analysis would be finding the input-output characteristic of the inverters shown in Fig. 1.1. The frequency response of a filter or amplifier would come from AC linear analysis. Small signal linear models are used for representing the active devices for doing AC analysis. The step response of a network would come under the category of transient analysis. In general, transient analysis involves finding the response of a network to excitations which have arbitrary variations as a function of time. It is the most time-consuming of the various analyses.

Whatever be the analysis done on a network, it is important to identify it with the corresponding numerical operations to be done. DC linear analysis, after proper formulation, reduces to the solution of a set of simultaneous linear equations. Non-linear DC analysis reduces to a set of simultaneous non-linear equations. Transient analysis generally involves solving a set of coupled non-linear differential equations. One can therefore infer the complexity of the analysis required. Perhaps the formulation is more important here than numerical analysis. Related to this is the issue of input description format of the circuit. For example, in order to analyse a DC linear network, node or mesh formulation must be made to first arrive at a set of simultaneous linear equations. Node or mesh formulation, in turn, is not possible unless the description of the circuit has been fed in some format. While numerical techniques are discussed in this book, the emphasis is on formulation.

1.4 Techniques for Large Digital Circuits

The larger ICs available today are mainly digital ICs. The semiconductor revolution has had the greatest impact on microprocessors and associated peripheral chips. Simulation of large digital circuits has therefore received a lot of attention. A digital circuit can be treated like any other circuit and its voltages and currents calculated. However, this very quickly becomes impractical for digital circuits having more than 50 to 100 gates. In any case, a simulation at the logic level is invariably needed at least as a preliminary check. Very large digital circuits can be (and are) simulated at the logic level without CPU times going out of hand. As mentioned in Section 1.1, the simulation can be at a behavioural level, RTL level or gate level. Often it is at more than one level with larger building blocks resulting from an hierarchical description [4].

A related activity is that of testing. Unless special care is taken during design, testing a VLSI/LSI chip may be an impossible task. The designer must provide a set of test input vectors so that the chip can be certifiable after fabrication. This in turn necessitates extensive fault simulation of the circuit. A lot of the simulation done at the logic level is for fault simulation [5].

More recently, a large variety of simulators have emerged for MOS digital circuits ([6], [7]). They calculate actual voltages as a function of time, i.e., they do transient analysis. Most of them exploit special features of MOS circuits and digital circuits. Concepts from gate level logic simulation have been used to speed up the execution by exploiting latency. These special purpose simulators mainly differ from circuit simulators like SPICE [3] in that they use relaxation methods to decouple the equations at the differential equation level, non-linear equation level or linear equation level [8].

1.5 Importance of Device Models

Circuit simulators like SPICE [3] would not be as useful as they are if they did not contain built-in models for common semiconductor devices like bipolar junction transistors (BJTs) and MOSFETs. Often the quality of simulation is decided by the accuracy of the models. Here, a compromise has to be struck between accuracy and speed of computation. A very detailed model would naturally slow down the program. Full fledged circuit simulators like SPICE [3] partially solve the problem by using a detailed model described in terms of a large number of parameters. By a proper choice of values for these parameters one can retain only those features of the model, important for the simulation one is doing. Timing simulators like MOTIS [6] use inherently simple models and the level of simulation may decide the complexity of the model.

Often, the user himself may have to build his own models. If a circuit containing many opamps is to be analysed, a model of the opamp is necessary [9]. Logic simulators may allow the user to define his own functional blocks. When using a circuit simulation package, the user usually has difficulty building his own non-linear model for a new device. The package or program would normally only allow non-linearities in the form of a polynomial. This is because derivatives are needed in analysing non-linear circuits as described in Chapter 4.

1.6 Special Network Elements

In addition to the network elements discussed so far, certain special elements have been invented to more conveniently describe active networks. Among these are gyrators, negative impedance converters (NICs), nullators and norators. While nullators and norators are idealised elements used to represent active networks gyrators and NICs are themselves realised using active elements like opamps. A gyrator can make a capacitor connected to its output terminals appear as an inductor. A NIC can do something similar and, in general, realise negative valued components. These special network elements are not considered any further in this book. Most practical circuits can be analysed and simulated without invoking these elements. The reader is directed to reference [10] for more details.

1.7 Summary

The stage has been set for the book in this first chapter. The need for simulation of circuits at various stages of design has been brought out, whether at the chip level or discrete component level. Networks and network elements have been classified as linear and non-linear. The various kinds of analyses normally done, DC (linear and non-linear), AC linear and transient linear and non-linear have been introduced. Circuit simulation may be done at many levels-behavioural, functional, logic or circuit level depending on the speed and accuracy required.

References

1. Y. Chu, Computer Organisation and Microprogramming, Chap. 1, Prentice Hall Inc., 1972.
2. F.J. Hill and G.R. Peterson, Digital Systems: Hardware Organisation and Design, Chap. 5, John Wiley and Sons, 1973.
3. L.W. Negel, SPICE 2: A Computer Program to Simulale Semiconductor Circuits, Ph.D. Thesis, University of California, Berkeley, May, 1978.
4. 'LAMP', Bell System Technical Journal, pp. 1442-1475, Oct. 1974.
5. M.A. Breuer and A.D. Friedman, Diagnosis and Reliable Design of Digital Systems, Chap. 4, Computer Science Press, 1976.
6. B.R. Chawla, H.K. Gummel and P. Kozak, MOTIS—an MOS Timing Simulator, IEEE Trans. Ccts. and Syst., vol. CAS-22, pp. 901-909, Dec. 1975.
7. S. Banerjee and R. Raghuram, MOSIMR: A Timing Simulator using Signal Propagation to Exploit Latency, First International Workshop on VLSI Design, Madras, India, Dec. 26-28, 1985.
8. A.R. Newton and A.L. Sangiovanni-Vincentelli, Relaxation-Based Electrical Simulation, IEEE Trans. CAD, vol. CAD-3, 4, pp. 308-331, Oct. 1984.
9. G.R. Boyle, B.M. Cohn, D.O. Pederson and J.E. Solomon, Macromodelling of IC Operational Amplifiers, IEEE J. Solid State Circuits, vol. SC-9, 6 pp. 353-363. Dec. 1974.
10. V.K. Aatre, Network Theory and Filter Design, pp 357-369, Wiley Eastern Ltd. 1980.

Problems

1. By applying the laws of homogeneity and additivity, check which of the following elements are linear.

   (a) A resistor with $I = I_0(e^{kV} - 1)$

   (b) A resistor with

   $$I = kV \quad \text{for } V \text{ greater than or equal to zero}$$
   $$I = 0 \quad \text{for } V \text{ less than zero}$$

(c) A capacitor whose value varies with time as $C = C_0 + C_1 \cos \omega t$ where C_0 and C_1 are positive constants and C_0 is greater than C_1.

2. Verify that the two models given for a pair of coupled coils with L_1, L_2 and M are valid.

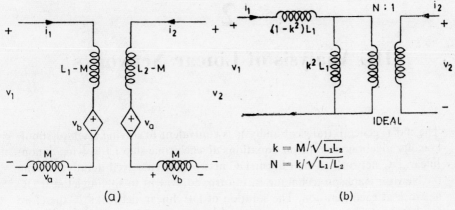

$$k = M/\sqrt{L_1 L_2}$$
$$N = k/\sqrt{L_1/L_2}$$

(a)   (b)

Fig. P1.2

3. Consider capacitors having Q as a linear function of v and inductors having λ as a linear function of i. Show that both forms of their $i-v$ relations (i.e. differential and integral forms) satisfy homogeneity and additivity.

4. In Fig. 1.6 find L_1'', L_2', M' and M'' in terms of L_1, L_2, M, N_1, N_2.

5. Given an example of a device (apart from the ones in the text) having a non-linear $i-v$ characteristics which is (a) strictly current controlled (b) strictly voltage controlled.

6. Inductors having magnetic materials as the core are non-linear because of the non-linear $B-H$ curve of the magnetic material. The $\lambda - i$ curve of such an inductor has been idealised as shown in Fig. P1.6. Find the inductance (λ/i) and effective inductance ($d\lambda/di$) as a function of i.

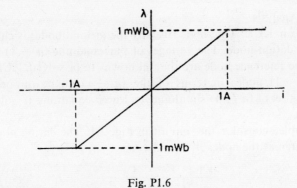

Fig. P1.6

7. The charge associated with the diffusion capacitance of a $p-n$ junction diode can be written as

$$Q = I_s \tau e^{V/V_T}$$

Find the current arising out of this capacitance for a given $\dfrac{dv}{dt}$.

2

DC Analysis of Linear Networks

The most general transient analysis is equivalent to solving a set of simultaneous algebraic non-linear equations at each time step i.e. solving a non-linear DC network. A non-linear DC network when solved iteratively using the Newton-Raphson technique is, in turn, equivalent to solving a linear DC network at each iteration. The solution of DC linear networks is therefore fundamental to any circuit simulation program. Formulation techniques for such networks are discussed in this section. The solution of the resulting equations is discussed in the next chapter.

2.1 Common Techniques

In principle, Kirchoff's current and voltage laws (KCL and KVL) are adequate to formulate the equations to analyse any circuit. However, if one writes arbitrary KCL and KVL equations, one may not have independent equations or the number of equations may not be adequate. Two basic techniques used are node and loop analysis. These are discussed first followed by some hybrid techniques.

2.1.1 Node Analysis
It is based on KCL. One node (usually the ground node) is chosen as the reference or datum node. The voltages of the remaining $(n - 1)$ nodes with respect to the reference node are the unknowns to be solved. KCL is written at these $(n - 1)$ nodes in terms of the $(n - 1)$ unknown node voltages. The resulting set of $(n - 1)$ simultaneous linear equations is solved to get the voltages.

For example, consider the circuit in Fig. 2.1. The datum node is 0 and KCL is written at the nodes A, B and C as

$$\frac{V_A - V_B}{R_1} + \frac{V_A - V_C}{R_4} = I_1$$

$$\frac{V_B - V_A}{R_1} + \frac{V_B}{R_2} - g_m(V_A - V_C) = 0$$

$$\frac{V_C}{R_3} + g_m(V_A - V_C) + \frac{V_C - V_A}{R_4} = 0$$

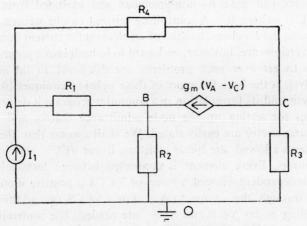

Fig. 2.1   Circuit for node analysis.

Rewriting in matrix form, one gets

$$\begin{bmatrix} \left(\dfrac{1}{R_1} + \dfrac{1}{R_4}\right) & \dfrac{-1}{R_1} & \dfrac{-1}{R_4} \\[2ex] \left(\dfrac{-1}{R_1} - g_m\right) & \left(\dfrac{1}{R_1} + \dfrac{1}{R_2}\right) & g_m \\[2ex] \left(g_m - \dfrac{1}{R_4}\right) & 0 & \left(\dfrac{1}{R_3} + \dfrac{1}{R_4} - g_m\right) \end{bmatrix} \begin{bmatrix} V_A \\[2ex] V_B \\[2ex] V_C \end{bmatrix} = \begin{bmatrix} I_1 \\[2ex] 0 \\[2ex] 0 \end{bmatrix}$$

This can be written in the form

$$[Y][V] = [I] \tag{2.1}$$

where

$\quad [Y] =$ node admittance matrix

$\quad [V] =$ vector of unknown voltages

$\quad [I] =$ equivalent current source vector

Formulation using node analysis leads to equations of the form of Eq. (2.1) above. [I] is called the equivalent current source vector because it may also contain elements having the dimensions of current but derived from independent voltage sources. Possible network elements in DC linear analysis are linear resistors and the four kinds of controlled sources. [Y] is constructed from the network elements while [I] is constructed from the independent sources or excitations. Node analysis is strictly applicable only to circuits containing resistors, VCCS' and independent current sources. If one of the elements at a node is a voltage source (independent or controlled) it is not possible to write the current through it. Similarly, if a controlling quantity is a current, node analysis does not work because the unknowns have to be voltages. Often, there are ways around it. If the voltage source has a series resistance one can convert it to its Norton equivalent. A grounded voltage source can have its non-grounded end excluded from the set of nodes KCL is written for. A controlling current can be written in terms of voltages using the *i–v* characteristics of an element the current flows through. These conversions are, however, awkward to be built into a program. Hybrid techniques to get over such problems are discussed in the next section. Node analysis is the basis for many of these hybrid techniques like modified node analysis and its formulation in a computer program is described below.

The rules for setting up the node admittance matrix and equivalent current source vector are easily stated. We shall assume that the only network elements allowed are linear resistors, linear VCCS' and independent current sources. Every element is connected between terminals $N+$ and $N-$. An independent current source or VCCS is positive if current flows from $N+$ through the source to $N-$. For a VCCS, two additional nodes, the controlling nodes $NC+$ and $NC-$, are needed. The controlling voltage is $[V(NC+) - V(NC-)]$. The ground or reference node is always given the number 0. The description of the circuit of Fig. 2.2 would be as follows:

Branch Number	$N+$	$N-$	Type	Value	$NC+$	$NC-$
1	1	2	R	1000	—	—
2	3	2	G	2	1	3
3	3	0	R	3000	—	—
4	2	0	R	4000	—	—
5	0	1	I	2	—	—
6	1	3	R	6000	—	—

The TYPE column indicates whether the element is a resistor (R), a VCCS (G) or an independent current source (I).

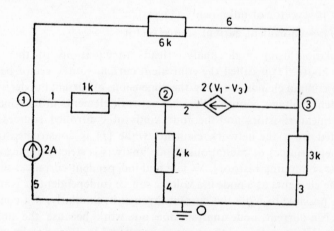

Fig. 2.2   Circuit for which node based input description has been given in text.

For the node-based description given above, the steps in arriving at the node admittance matrix $[Y]$ and equivalent current source $[I]$ vector are as follows:

1. Initialise all $Y(i, j)$ and $I(i)$ to zero.
2. For every resistor R between nodes $N+$ and $N- (N+, N - \neq 0)$

$$Y(N+, N+) = Y(N+, N+) + \frac{1}{R}$$

$$Y(N--, N--) = Y(N-, N-) + \frac{1}{R}$$

$$Y(N-, N+) = Y(N-, N+) - \frac{1}{R}$$

$$Y(N+, N-) = Y(N+, N-) - \frac{1}{R}$$

3. For every resistor R between node $N(N$ can be $N+$ or $N-)$ and 0.

$$Y(N, N) = Y(N, N) + \frac{1}{R}$$

4. For every VCCS G between $N+$ and $N-$ with controlling nodes $NC+$ and $NC-$

$$Y(N+, NC+) = Y(N+, NC+) + G$$
$$Y(N+, NC-) = Y(N+, NC-) - G$$
$$Y(N-, NC+) = Y(N-, NC+) - G$$
$$Y(N-, NC-) = Y(N-, NC-) + G$$

If any of these nodes $N+$, $N-$, $NC+$, $NC-$ is 0, is the corresponding row and column entries above are omitted.

5. For every independent current source I_s from $N+$ to $N-$

$$I(N+) = I(N+) - I_s$$
$$I(N-) = I(N-) + I_s$$

We now get a set of $(n - 1)$ simultaneous linear equations. These equations are solved using the various techniques described in the next chapter. The above rules are simplified if the ground node is treated like any other node. In the end the row and column corresponding to this node can be deleted.

2.1.2 Loop Analysis

This is a dual technique and is based on writing KVL for a set of independent loops, where currents are the unknowns. If the network has b branches and n nodes, loop analysis leads to a set of $(b - n + 1)$ linear equations. In order to identify the independent loops a tree of the network graph is formed. The tree is a connected subgraph which contains all the nodes but no loops or closed paths. Many trees are possible for a graph. The network of Fig. 2.1 can be represented by the graph of Fig. 2.3(a). The branches of the network graph not included in the tree are called links or chords. By introducing one chord at a time, different loops can be formed. Loops corresponding to two different trees are shown in Fig. 2.3(b) and Fig. 2.3(c). If KVL equations are written for the set of loops corresponding to any tree, one arrives at a set of independent loop equations. Often it is more convenient to do mesh analysis, which can be viewed as a special case of loop analysis. Here the tree is so chosen that none of the loops have closed paths inside them in the original network. Figure 2.3(c) corresponds to such a choice. (It may not always be possible to choose such a tree). The main advantage in mesh analysis is that the windows or meshes can be identified by inspection and one does not need to construct a tree. Its major limitation is that it is applicable only to planar networks. A planar network has a graph which can be drawn on paper without any two branches intersecting each other.

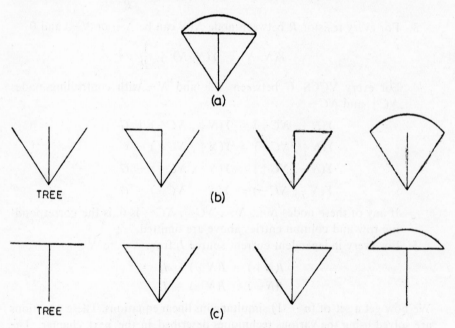

Fig. 2.3   (a) Network graph for circuit of Fig. 2.1
(b) A possible tree and associated loops
(c) A tree which leads to windows or meshes as loops.

As an example consider the network of Fig. 2.4. The meshes are identified by inspection and the mesh currents chosen as I_1, I_2 and I_3. Mesh equations give

$$I_1 R_1 + (I_1 - I_3)R_3 + (I_1 - I_2)R_4 = E_1$$
$$(I_2 - I_1)R_4 + (I_2 - I_3)R_5 + I_2 R_2 = -E_2$$
$$I_3 R_6 + (I_3 - I_2)R_5 + (I_3 - I_1)R_3 = 0$$

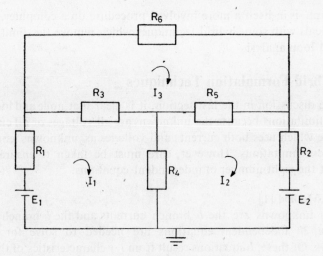

Fig. 2.4 Circuit for mesh analysis.

These can be rewritten as:

$$
\begin{bmatrix}
(R_1 + R_3 + R_4) & -R_4 & -R_3 \\
-R_4 & (R_4 + R_5 + R_2) & -R_5 \\
-R_3 & -R_5 & (R_6 + R_5 + R_3)
\end{bmatrix}
\begin{bmatrix}
I_1 \\
I_2 \\
I_3
\end{bmatrix}
$$

$$
=
\begin{bmatrix}
E_1 \\
-E_2 \\
0
\end{bmatrix}
$$

This is of the form

$$[Z][I] = [E] \tag{2.2}$$

where

$[Z]$ = loop impedance matrix
$[I]$ = vector of unknown loop currents
$[E]$ = equivalent voltage source vector

Loop analysis has a dual set of limitations when compared with node analysis. Current sources, independent or controlled, cannot be handled directly. Controlling quantities have to be currents as they are the unknowns. Of the four controlled sources, only the CCVS can be directly analysed by loop analysis.

Node and loop analysis appear equally applicable for analysing a network. However, most circuit analysis packages use node analysis or a variation of it. This is because a node based description is most convenient. The node admittance matrix and equivalent current source vector are directly constructed from such a description. On the other hand, identifying the loops of a graph is in itself a more involved procedure on a computer. The next section deals with some hybrid techniques which remove the limitations of node and loop analysis.

2.2 Hybrid Formulation Techniques

From the discussion in the last section, it is clear that node and loop analysis have limitations because the unknowns are all voltages or all currents. A technique which uses both currents and voltages as unknowns could overcome these limitations. However, care must be taken to ensure that one arrives at the right number of independent equations.

2.2.1 2b Method [1]

Here the unknowns are the b branch currents and the b branch voltages. Therefore, $2b$ independent equations are needed to solve for these $2b$ unknowns. Of these, b equations result from i-v characteristics of the various branch elements. These are called the branch constitutive relations. Then $(n - 1)$ node equations are written in terms of the branch currents. Finally, $(b - n + 1)$ loop equations are written in terms of the branch voltages. It can be shown that these $2b$ equations are independent. While the b branch constitutive equations depend on the network elements, the remaining $(n - 1)$ node equations and the $(b - n + 1)$ loop equations are strictly a function of the topology.

For the network of Fig. 2.5(a), the following equations would result:
1. Branch constitutive relations

$$V_1 = -E_0$$
$$V_2 = R_2 I_2$$
$$V_3 = R_3 I_3$$
$$V_4 = R_4 I_4$$
$$I_5 = \alpha I_2$$
$$V_6 = R_6 I_6$$

2. $(n - 1)$ node equations from Fig. 2.5(b)

$$-I_1 + I_2 + I_6 = 0$$
$$-I_2 + I_3 + I_4 = 0$$
$$-I_5 - I_4 - I_6 = 0$$

3. $(b - n + 1)$ loop equations from Fig. 2.5(b)

$$V_1 + V_2 + V_3 = 0$$
$$V_4 - V_5 - V_3 = 0$$
$$V_2 + V_4 - V_6 = 0$$

The number of unknowns in the $2b$ method is considerably larger than in node or loop analysis. However, the matrices obtained are sparse and this can be exploited as discussed in Chapter 3. Writing the loop equations again involves identifying the loops and this may be difficult. The Tableau method overcomes this problem.

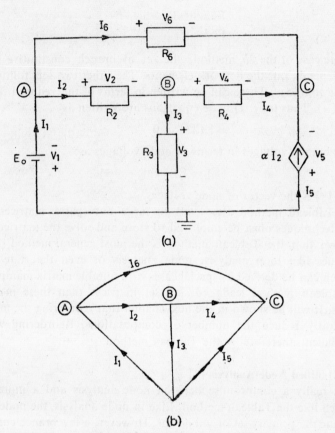

Fig. 2.5   (a) Circuit analysed by 2b method and Tableau method,
          (b) Corresponding graph, current unknowns and
              node voltages.

2.2.2 Tableau Method [2]

This has an even larger number of unknowns and equations. The unknowns are the b branch currents, the b branch voltages and the $(n-1)$ node voltages. In all, there are $(2b+n-1)$ unknowns and the following equations are written:

1. b branch constitutive relations
2. $(n-1)$ node equations in terms of the branch currents
3. b equations giving the branch voltages in terms of the node voltages.

The last set of equations is different from the 2b method. For the network of Fig. 2.5, these equations are

$$V_1 = -V_A$$

$$V_2 = V_A - V_B$$
$$V_3 = V_B$$
$$V_4 = V_B - V_C$$
$$V_5 = -V_C$$
$$V_6 = V_A - V_C$$

As in the case of the $2b$ method, the set of branch constitutive relations alone depends on the network elements. The other two sets follow strictly from the topology. These can be written in terms of the reduced incidence matrix $[A]$. Thus $(n-1)$ node equations are written as:

$$[A][I] = 0 \qquad (2.3)$$

and the branch voltages in terms of node voltages as:

$$[V] = [A^t][V^n]$$

where $[V^n]$ is the vector of node voltages.

The Tableau method also leads to large but sparse matrices. Sparse matrix techniques must be employed to store and solve the matrices. It can be shown that the Tableau method is the most general method [3]. Other techniques like loop analysis, node analysis or even the state variable approach can be derived as special cases with suitable matrix manipulations. The Tableau method leads to sparser matrices than these methods of analysis. It will be shown in the next chapter that reordering the matrix can significantly reduce the number of computations. Reordering would be most efficient, therefore, in the Tableau method.

2.2.3 Modified Node Analysis [4]

This is really a compromise between node analysis and a more general approach like the Tableau method. Like in node analysis, the node voltages constitute the primary set of unknowns. However, a few branch currents are also chosen as unknowns to take care of the element not directly handled by node analysis. These are

1. Current through any voltage source, independent or controlled.
2. Any controlling branch current as for a CCCS or CCVS.
3. Any branch current which is to be explicitly determined. This will be necessary, for example, in current controlled non-linear elements as will be seen in Chapter 4.

Once the unknowns are chosen, the node equations are written at all the nodes except the datum node. In writing these equations, wherever possible, the branch currents chosen as unknowns should be used. Finally, a branch constitutive relation is written for each branch for which the current is chosen as one of the unknowns. These branch constitutive relations supply the extra equations needed for the extra branch current unknowns. For the circuit shown in Fig. 2.6, the primary unknowns would be the node voltages V_A, V_B, V_C and V_D. In addition, the currents through the voltage sources,

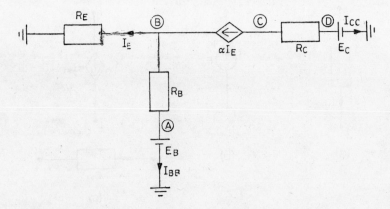

Fig. 2.6 Circuit for modified node analysis. This corresponds to a transistor in the normally active region.

I_{BB} and I_{CC}, as well as the controlling current I_E would be chosen as unknowns. The node equations at A, B, C and D are

$$\frac{V_A - V_B}{R_B} + I_{BB} = 0$$

$$I_E - \alpha I_E + \frac{V_B - V_A}{R_B} = 0$$

$$\alpha I_E + \frac{V_C - V_D}{R_C} = 0$$

$$\frac{V_D - V_C}{R_C} + I_{CC} = 0$$

The branch constitutive relations are

$$V_A = E_B$$

$$V_D = E_C$$

$$\frac{V_B}{R_E} - I_E = 0$$

The terms of the modified node admittance matrix $[Y]$ and equivalent current source vector $[I]$ corresponding to various circuit elements are given below. Figure 2.7 gives the current and voltage polarity conventions adopted for the various elements. The controlling branch for a CCCS and a CCVS has been taken as a linear resistor. Programs like SPICE [6] and the one given at the end of the chapter require that it be an independent voltage source. If no source exists, as for example in Fig. 2.8(b), a $0V$ voltage source can be added in series.

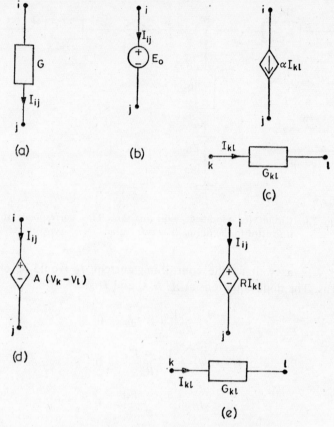

Fig. 2.7 Current and voltage conventions adopted for defining matrix elements for modified node analysis:

(a) Linear conductor through which current is needed
(b) Independent voltage source
(c) Linear CCCS
(d) Linear VCVS
(e) Linear CVCS. For the CCVS and CCCS, the controlling branch with nodes k and l is also shown.

(a) Linear Conductor G through which current is to be determined

	[Y]			[I]
	V_i ...	V_J ...	I_{ij}	
Node i			$+1$	—
Node j			-1	—
Branch relation	G ...	$-G$	-1	—

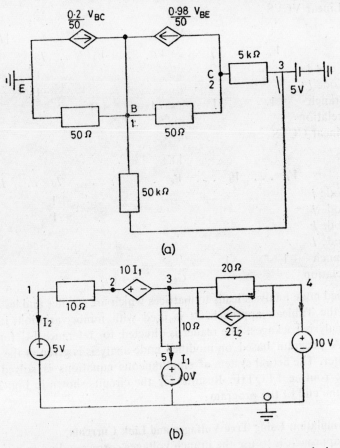

Fig. 2.8 Circuits analysed using the modified node analysis
program given.

(b) Independent voltage source E_0

	[Y]				[I]
	V_i	...	V_j	...	I_{ij}
Node i					$+1$	$-$
Node j					-1	$-$
Branch relation	$+1$		-1			E_0

(c) Linear CCCS

	[Y]			[I]
	V_k	...	V_1	...	I_{k1}
Node i					α	$-$
Node j					$-\alpha$	$-$
Node k					$+1$	$-$
Node l					-1	$-$
Branch relation	G_{k1}		$-G_{k1}$			$-$

(d) Linear VCVS

	[Y]					[I]
	V_i ...	V_j ...	V_k ...	V_l ...	I_{ij}	
Node i					+1	—
Node j					−1	—
Branch relation	+1	−1	$-A$	$+A$		—

(e) Linear CCVS

	[Y]						[I]
	V_i ...	V_j ...	V_k ...	V_l ...	I_{ij} ...	I_{ki}	—
Node i					+1		—
Node j					−1		—
Node k						+1	—
Node l						−1	—
Branch	+1	−1				$-R$	—
relation			G_{kl}	$-G_{kl}$		−1	—

Modified node analysis leads to matrices which are smaller and less sparse than in the Tableau approach. It is used with minor variations in many circuit analysis packages. The reader is directed to reference [4] for more details. A program based on modified node analysis is given at the end of this chapter. The actual system of simultaneous equations is solved using an IMSL routine LEQTIF. Results for the circuits shown in Fig. 2.8 are given at the end of the program.

2.2.4 Formulation Using Tree Voltages and Link Currents

Another approach is to use the branch voltages corresponding to a tree and the corresponding link currents as unknowns [6]. The branches of a tree are referred to as twigs. It can be shown that the twig currents and the link voltages can be expressed in terms of twig voltages, V_t, and link currents I_l. Corresponding to the choice of tree, one can write KVL for the loops and KCL for the cutsets. Finally, one gets equations of the form

$$\begin{bmatrix} Y_t & H_{21} \\ H_{12} & Z_l \end{bmatrix} \begin{bmatrix} V_t \\ I_l \end{bmatrix} = \begin{bmatrix} I_s \\ V_s \end{bmatrix}$$

where I_s and V_s are derived from the independent current and voltage sources. The choice of tree should be such that voltage sources (dependent or independent) are included in the links and current sources (dependent or independent) in the twigs. A controlling branch whose current is the controlling quantity is included among the links. A controlling branch whose voltage is the controlling quantity is included in the twigs.

This method involves finding trees, loops, cutsets, besides considerable matrix manipulations. The discussion in the rest of the book is based mainly on modified mode analysis.

2.3 Summary

Various techniques for the formulation of equations for DC linear circuits
have been discussed in this chapter. Both currents and voltages have to be
chosen as unknowns in order to analyse such circuits in general. The most
general method is the Tableau method which leads to a large number of
equations. Modified node analysis is a technique more commonly used.
Elements not handled by node analysis are handled by choosing associated
currents as unknowns. Modified node analysis leads to sparser matrices than
node analysis, but the Tableau method leads to even sparser matrices.

```
C         PROGRAM FOR DC ANALYSIS USING MODIFIED NODE VOLTAGE
C         FORMULATION
C         GROUND OR REFERENCE NODE IS TAKEN AS 0
          DIMENSION A(20,20), B(20), KC(10), KE(4),WA(1000),C(20,20)
C         A IS THE MODIFIED NODE ADMITTANCE MATRIX. B IS THE EQUIVALENT
C         CURRENT SOURCE VECTOR AND CONTAINS THE SOLUTION AT THE END.
C         OTHER ARRAYS ARE FOR TEMPORARY STORAGE.
          OPEN(UNIT=20,DEVICE='DSK',FILE='MNA.IN')
          OPEN(UNIT=22,DEVICE='DSK',FILE='MNA.OUT',ACCESS='SEQOUT')
          READ(20,7), (KC(I),I=1,7)
7         FORMAT(7A1)
C         KC DENOTES HOW THE VARIOUS ELEMENTS ARE REFERRED TO.
C         INITIALISE ARRAYS
          DO 9 I=1,20
          B(I)=0.0
          DO 9 J=1,20
          A(I,J)=0.0
9         CONTINUE
          READ(20,2)ND,N
2         FORMAT(2I3)
C         ND IS THE NUMBER OF BRANCHES. N IS THE NUMBER OF NODES
C         INCLUDING THE GROUND OR REFERENCE NODE.
          IN=N
C         IN IS AN INDEX UPCATING UNKNOWNS.
          DO 200 I=1,ND
          READ(20,25),NA,NP,NM,VAL,NCP,NCM
25        FORMAT(A1,2I3,F10.4,2I3)
C         ALL NODE NUMBERS ARE INCREASED BY 1 TO INCLUDE REFERENCE NODE
          NP=NP+1
          NM=NM+1
          NCP=NCP+1
          NCM=NCM+1
C         NA IS THE CHARACTER INDICATING THE ELEMENT BEING READ
C         NA IS COMPARED WITH VALUES IN KC TO IDENTIFY ELEMENT
          IF (NA.EQ.KC(1)) GOTO 30
          IF (NA.EQ.KC(2)) GOTO 50
          IF (NA.EQ.KC(3)) GOTO 70
          IF (NA.EQ.KC(4)) GOTO 90
          IF (NA.EQ.KC(5)) GOTO 110
          IF (NA.EQ.KC(6)) GOTO 130
          IF (NA.EQ.KC(7)) GOTO 150
C         IN THE FOLLOWING NP IS THE +NODE AND NM IS THE -NODE
C         NCP AND NCM ARE + AND - NODES OF CONTROLLING ELEMENT.
30        A(NP,NP)=A(NP,NP) + 1.0/VAL
          A(NP,NM)=A(NP,NM) -1.0/VAL
          A(NM,NP)=A(NM,NP) -1.0/VAL
          A(NM,NM)=A(NM,NM) + 1.0/VAL
          GOTO 200
C         VCCS
50        A(NP,NCP)=A(NP,NCP)+VAL
          A(NM,NCM)=A(NM,NCM)+VAL
          A(NP,NCM)=A(NP,NCM)-VAL
          A(NM,NCP)=A(NM,NCP)-VAL
          GOTO 200
C         VCVS: CONTROLLING BRANCH MUST BE RESISTOR OR VCCS.
70        IN=IN+1
85        A(IN,NP)=A(IN,NP)+1 0
          A(IN,NM)=A(IN,NM)-1 0
```

```
              A(IN,NCP)=A(IN,NCP)-VAL
              A(IN,NCM)=A(IN,NCM)+VAL
              A(NP,IN)=A(NP,IN)+1.0
              A(NM,IN)=A(NM,IN)-1.0
              GOTO 200
C             CCCS: CONTROLLING CURRENT MUST BE CURRENT THROUGH IND. VOLT. SRC
C             ONLY ONE CCCS/CCVS CAN BE CONTROLLED BY  ONE CONTROLLING CURRENT
90            IN=IN+1
105           A(NP,IN)=A(NP,IN)+VAL
              A(NM,IN)=A(NM,IN)-VAL
              A(NCP,IN)=A(NCP,IN)+1.0
              A(NCM,IN)=A(NCM,IN)-1.0
              A(IN,NCP)=A(IN,NCP)+1.0
              A(IN,NCM)=A(IN,NCM)-1.0
              GOTO 200
C             CCVS: CONTROLLING CURRENT MUST BE CURRENT THROUGH IND. VOLT. SRC
C             ONLY ONE CCCS/CCVS CAN BE CONTROLLED BY ONE CONTROLLING CURRENT
110           IN=IN+1
125           IN2=IN+1
              A(IN,NP)=A(IN,NP)+1.0
              A(IN,NM)=A(IN,NM)-1.0
              A(IN,IN2)=A(IN,IN2)-VAL
              A(NM,IN)=A(NM,IN)-1.0
              A(NP,IN)=A(NP,IN)+1.0
              A(NCP,IN2)=A(NCP,IN2)+1.0
              A(NCM,IN2)=A(NCM,IN2)-1.0
              A(IN2,NCP)=A(IN2,NCP)+1.0
              A(IN2,NCM)=A(IN2,NCM)-1.0
              IN=IN2
              GOTO 200
C             INDEPENDENT VOLTAGE SOURCE
130           IN=IN+1
C             IF VOLTAGE SOURCE IS CONTROLLING BRANCH JUST UPDATE
C             EQUIVALENT CURRENT SOURCE VECTOR.
143           DO 145 L=N+1,IN
              IF(A(L,NP).EQ.0.0) GOTO 145
              IF(A(L,NM).EQ.0.0) GOTO 145
              ITEMP=L
              IN=IN-1
              GOTO 149
145           CONTINUE
148           A(IN,NP)=A(IN,NP)+1.0
              A(IN,NM)=A(IN,NM)-1.0
              A(NP,IN)=A(NP,IN)+1.0
              A(NM,IN)=A(NM,IN)-1.0
              B(IN)=B(IN)+VAL
              GOTO 200
149           B(ITEMP)=B(ITEMP)+VAL
              GOTO 200
C             INDEPENDENT CURRENT SOURCE. CURRENT FLOWS FROM NP
C             THROUGH SOURCE TO NM.
150           B(NP)=B(NP)-VAL
              B(NM)=B(NM)+VAL
200           CONTINUE
245           WRITE(22,31)
31            FORMAT(12X,' ** NODE ADMITTANCE MATRIX ** '/12X,
             1'_____'////)
              DO 79 I=2,IN
```

```
79        WRITE(22,33),(A(I,J),J=2,IN)
33        FORMAT(50(E12.5,2X))
          WRITE(22,47)
47        FORMAT(///,12X' * EQUIVALENT CURRENT SOURCE VECTOR * '/12X,
          1'_____'///)
          WRITE(22,58),(B(I),I=2,IN)
58        FORMAT(12(E12.5,2X))
C         THE NODE ADMITTANCE MATRIX IS SOLVED USING THE IMSL ROUTINE
C         FOR SOLVING LINEAR EQUATIONS.
C         COLUMNS AND ROWS CORRESPONDING TO THE REFERENCE NODE ARE DELETED
          DO 38 I=1,IN-1
          B(I)=B(I+1)
          DO 38 J=1,IN-1
38        C(I,J)=A(I+1,J+1)
          K=IN-1
          CALL LEQT1F(C,1,K,20,B,2,WA,IER)
          NU=N-1
          IU=IN-N
          WRITE(22,16),NU
16        FORMAT(//12X,'** DC ANALYSIS OF GIVEN CIRCUIT **'/12X,
          1'_____'//20X,
          1'** ',I2,' NODE VOLTAGES  **'/20X,
          1'_____')
          DO 61 I=1,NU
61        WRITE(22,62),I,I,B(I)
62        FORMAT(12(I2,' ',15X,'V(',I2,') = ',E12.5))
          WRITE(22,86),IU
86        FORMAT(//20X,'** ',I2,' SOURCE CURRENTS  **'/
          120X,'_____')
          IF (IU.EQ.0) STOP
          DO 97 I=N,IN-1
          IF=I-N+1
97        WRITE(22,99),I,IF,B(I)
99        FORMAT(12(I2,' ',15X,'I(',I2,') = ',E12.5))
          CLOSE(UNIT=20)
          CLOSE(UNIT=22)
          STOP
          END
```

** NODE ADMITTANCE MATRIX **

```
 0.16420E-01    -0.16000E-01    -0.20000E-04     0.00000E+00
-0.40000E-03     0.20200E-01    -0.20000E-03     0.00000E+00
-0.20000E-04    -0.20000E-03     0.22000E-03     0.10000E+01
 0.00000E+00     0.00000E+00     0.10000E+01     0.00000E+00
```

* EQUIVALENT CURRENT SOURCE VECTOR *

```
0.00000E+00     0.00000E+00     0.00000E+00     0.50000E+01
```

** DC ANALYSIS OF GIVEN CIRCUIT **

** 3 NODE VOLTAGES **

```
1              V( 1) =    0.55398E-01
2              V( 2) =    0.50602E-01
3              V( 3) =    0.50000E+01
```

** 1 SOURCE CURRENTS **

```
4              I( 1) =   -0.10888E-02
```

** NODE ADMITTANCE MATRIX **

```
0.10000E+00  -0.10000E+00   0.00000E+00   0.00000E+00   0.00000E+00   0.10000E+01   0.00000E+00   0.00000E+00
-0.10000E+00  -0.10000E+00   0.00000E+00   0.10000E+01   0.10000E+01   0.00000E+00   0.00000E+00   0.00000E+00
0.00000E+00   0.00000E+00   0.15000E+00  -0.10000E+00  -0.20000E+01  -0.10000E+00   0.10000E+01   0.00000E+00
0.00000E+00   0.00000E+00  -0.50000E-01  -0.10000E+00  -0.20000E+01   0.00000E+00   0.00000E+00   0.10000E+01
0.10000E+01   0.00000E+00   0.00000E+00   0.10000E+01   0.00000E+00   0.00000E+00   0.00000E+00   0.00000E+00
0.00000E+00   0.10000E+01   0.00000E+00   0.00000E+00   0.00000E+00   0.00000E+00  -0.10000E+02   0.00000E+00
0.00000E+00   0.00000E+00   0.10000E+01   0.00000E+00   0.00000E+00   0.00000E+00   0.00000E+00   0.00000E+00
0.00000E+00   0.00000E+00   0.00000E+00   0.10000E+01   0.00000E+00   0.00000E+00   0.00000E+00   0.10000E+03
```

* EQUIVALENT CURRENT SOURCE VECTOR *

```
0.00000E+00   0.00000E+00   0.00000E+00   0.00000E+00   0.50000E+01   0.00000E+00   0.00000E+00   0.10000E+02
```

** DC ANALYSIS OF GIVEN CIRCUIT **

**   5   NODE VOLTAGES   **

```
V( 1) =   0.50000E+01
V( 2) =   0.00000E+00
V( 3) =   0.00000E+00
V( 4) =   0.10000E+02
V( 5) =   0.00000E+00
```

**   4   SOURCE CURRENTS   **

```
I( 1) =  -0.50000E+00
I( 2) =   0.50000E+00
I( 3) =   0.00000E+00
I( 4) =   0.50000E+00
```

References

1. R.E. Scott, Linear Circuits, Part 1, Time Domain Analysis, p. 49, Addison-Wesley Publishing Co., 1960.
2. C.D. Hachtel, R.R. Brayton and F.G. Gustavson, The Sparse Tableau Approach to Network Analysis and Design, IEEE Trans. Circuit Theory, vol. CT-18, pp. 101-103, Jan., 1971.
3. L.O. Chua and P. Lin, Computer—Aided Analysis of Electronic Circuits, p. 674, Prentice-Hall, Inc., 1975.
4. C.W. Ho, A.E. Ruehli and P.A. Brennan, The Modified Nodal Approach to Network Analysis, IEEE Trans. Ccts. and Syst., vol. CAS-22, pp. 504-509, June 1975.
5. L.W. Negel, SPICE 2: A Computer Program to Simulate Semiconductor Circuits, Ph.D. Thesis, University of California, Berkeley, May, 1978.
6. N. Balabanian and T.A. Bickart, Electrical Network Theory, pp. 131-139, John Wiley & Sons, 1979.

Problems

1. Node analysis of the network shown gave the following set of equations. Find g_m and the controlling branch x for the VCCS between A and C.

$$\begin{bmatrix} 0 & 1 & 0 \\ -1 & 3 & -1 \\ 2 & -3 & 2 \end{bmatrix} \begin{bmatrix} V_A \\ V_B \\ V_C \end{bmatrix} = \begin{bmatrix} 2 \\ 0 \\ 0 \end{bmatrix}$$

What is abnormal about this network and its node admittance matrix?

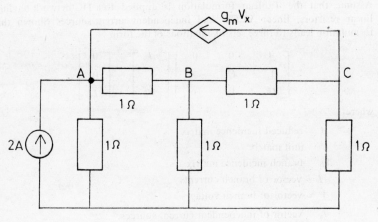

Fig. P 2.1

2. For the graph shown, find the tree which gives rise to loops which are all meshes.
3. For the networks shown in Fig. 2.8 find the quantities to be chosen as unknowns and write down the equations in matrix form. Consider the following hybrid techniques of analysis (a) Modified node analysis (b) Tableau method (c) 2b method.
4. Apply Tableau formulation to the network shown in Fig. 2.6. Compare the number of unknowns with that for modified node analysis.
5. Repeat Problem 3 for the circuit shown in Fig. P. 2.5.

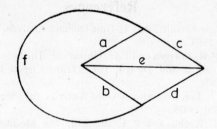

Fig. P 2.2

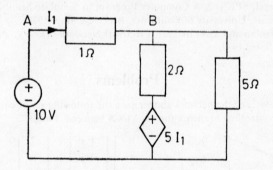

Fig. P 2.5

6. Assume that the Tableau formulation is applied to a DC network having only linear resistors, linear VCCS' and independent current sources. Shown that the formulation leads to three sets of equations of the form

$$\begin{bmatrix} A & 0 & 0 \\ 0 & U & -A^t \\ U & -Y_b & 0 \end{bmatrix} \begin{bmatrix} I \\ V \\ V_n \end{bmatrix} = \begin{bmatrix} 0 \\ 0 \\ I_s \end{bmatrix}$$

where

A = reduced incidence matrix

U = unit matrix

Y_b = branch incidence matrix

I = vector of branch currents

V = vector of branch voltages

I_s = vector of independent current sources

V_n = vector of node voltages

7. From the equations in Problem 6 derive the equation for node analysis as

$$[AY_bA^t][V_n] = [-AI_s]$$

Therefore $[AY_bA^t]$ = node admittance matrix

$$[-AI_s] = \text{equivalent current source vector}$$

8. Take any circuit you know with 10 or more nodes. See that any formulation leads to sparse matrices i.e. matrices with a large number of element values being zero.

Programming Assignments

1. Using the rules given in Section 2.1.1 write a program to analyse DC networks containing linear resistors, linear VCCS' and independent current sources. Assume the existence of a subroutine/procedure to solve simultaneous linear equations. (e.g. LEQT1F and LEQT2F of IMSL).
2. Modify the program in programming assignment 1 to include grounded independent voltage sources.
3. The program given for modified node analysis has the following limitations:
   (a) All controlling currents must be through an independent voltage source
   (b) Only one CCCS or CCVS can be controlled by a controlling current.
   Try to see why these limitations make programming easier. How would you get over these limitations?
4. Write a program for Tableau formulation to handle linear resistors, all 4 kinds of linear controlled sources and independent current and voltage sources. Use at library routine to solve the resulting simultaneous linear equations as before, but try to find one for sparse matrices.

3
Solution of Simultaneous Linear Algebraic Equations

It was seen in the last chapter that DC linear analysis finally boils down to solving a system of simultaneous linear equations. As will be seen in succeeding chapters, the same is true of non-linear DC and transient analyses. A routine for solving simultaneous linear equations forms the core of any circuit analysis program. Further, various hybrid techniques like Tableau method and modified node analysis lead to sparse matrices. In fact, for a circuit with more than about 20 nodes any formulation leads to sparse matrices. It is important that this sparsity be exploited. This is done partly in the storage and partly in ensuring that the sparsity is not destroyed during the solution.

The system of equations to be solved can be written as:

$$[A][x] = [B] \tag{3.1}$$

where,

$[A] = (n \times n)$ matrix like the node admittance matrix
$[x] = (n \times 1)$ vector of unknowns to be solved for
$[B] = (n \times 1)$ known vector.

Although a variety of techniques exist for solving these equations, many common ones can be ruled out. Cramer's rule involves finding $(n + 1)$ determinants. Evaluating each one of which is of the same order of complexity as solving the system of equations itself by other methods. It is grossly inefficient when compared to Gaussian elimination which requires $0(n^3/3)$ operations. Finding the inverse requires $0(n^3)$ operations and is also not as efficient as Gaussian elimination.

3.1 Gaussian Elimination

This is a very standard and popular method for solving simultaneous linear equations. For a general matrix, it is the fastest method. Solution proceeds in two stages

(a) Forward Reduction
(b) Back Substitution

In Forward Reduction, $[A]$ is reduced to an upper triangular matrix with all diagonal elements equal to 1. The elements a_{21} to a_{n1} are first made zero, then a_{32} to a_{n2}. then a_{43} to a_{n3} and so on. In order to make $a_{21} = 0$, for example, the first row of $[A]$ is multiplied by $(-a_{21}/a_{11})$ and added to the second row. The column vector $[B]$ is added as an extra column to $[A]$. This is called the augmented matrix. Using the augmented corresponding changes are automatically made in $[B]$ as well. The steps required to make a_{21} to a_{n1} equal to zero are

1. Divide first row of augmented matrix $[A]$ by a_{11}
2. For $i = 2$ to n multiply the first row by $-a_{11}$ and add to ith row.

In doing these operations the element a_{11} is said to be the pivot. Next the submatrix, with the first row and column of the augmented matrix $[A]$ deleted, is considered. The element a_{22} (modified by operations using a_{11} as pivot) is chosen as pivot and elements a_{32} to a_{n2} are made zero. This goes on till a_{nn-1} has been made zero using $a_{n-1\ n-1}$ as pivot. Consider the following set of equations:

$$\begin{bmatrix} 3 & 1 & 1 \\ 1 & 1 & 1 \\ 2 & 1 & 2 \end{bmatrix} \begin{bmatrix} x_1 \\ x_2 \\ x_3 \end{bmatrix} = \begin{bmatrix} 8 \\ 6 \\ 10 \end{bmatrix}$$

The augmented matrix at various stages of Forward Reduction is as given below:

$$\begin{bmatrix} 3 & 1 & 1 & 8 \\ 1 & 1 & 1 & 6 \\ 2 & 1 & 2 & 10 \end{bmatrix} \begin{bmatrix} 1 & \dfrac{1}{3} & \dfrac{1}{3} & \dfrac{8}{3} \\ 0 & \dfrac{2}{3} & \dfrac{2}{3} & \dfrac{10}{3} \\ 0 & \dfrac{1}{3} & \dfrac{4}{3} & \dfrac{14}{3} \end{bmatrix}$$

$$\begin{bmatrix} 1 & \dfrac{1}{3} & \dfrac{1}{3} & \dfrac{8}{3} \\ 0 & 1 & 1 & 5 \\ 0 & 0 & 1 & 3 \end{bmatrix}$$

The final form above can be rewritten as:

$$\begin{bmatrix} 1 & \dfrac{1}{3} & \dfrac{1}{3} \\ 0 & 1 & 1 \\ 0 & 0 & 1 \end{bmatrix} \begin{bmatrix} x_1 \\ x_2 \\ x_3 \end{bmatrix} = \begin{bmatrix} \dfrac{8}{3} \\ 5 \\ 3 \end{bmatrix}$$

Once the equations are brought to the upper triangular form by Forward Reduction, the solution is easy. The variable x_n is directly given by the last equation. Substituting for x_n in the last but one equation gives x_{n-1} and so on. This process is called Back Substitution.

It is important to calculate the number of operations in Gaussian elimination so that it may be compared with other methods. For this purpose only multiplications and divisions are considered as operations. It is assumed that additions and substractions take a negligible amount of time on a computer compared with multiplication and division. Consider the first column of the augmented matrix $[A]$ in doing Forward Reduction. Division by a_{11} takes n operations. For each i from 2 to n the first row is to be multiplied by $-a_{i1}$ (and then added to ith row). This takes n operations for each i. The total number of operations with a_{11} as pivot is therefore

$$n + n(n - 1) = n^2 \text{ operations}$$

The total number of operations in Forward Reduction is therefore

$$n^2 + (n - 1)^2 + (n - 2)^2 + \ldots 2^2 + 1^2 = \sum_{i=1}^{n} i^2$$

$$= \frac{n^3}{3} + \frac{n^2}{2} + \frac{n}{6}$$

In doing Back Substitution, x_n is given directly. Finding x_{n-1} requires one operation, x_{n-2} two operations, etc. Finally, x_1 requires $(n - 1)$ operations. Total number of operations in Back Substitution is

$$1 + 2 + 3 + \ldots n - 1 = \sum_{i=1}^{n-1} i = \frac{n(n - 1)}{2}$$

The total number for Gaussian elimination is therefore

$$\left(\frac{n^3}{3} + \frac{n^2}{2} + \frac{n}{6} \right) + \frac{n(n - 1)}{2} = \frac{n^3}{3} + n^2 - \frac{n}{3}$$

For a general matrix no method is superior to Gaussian elimination in the number of operations.

One point glossed over so far is what happens if the pivot element is zero at any stage during Forward Reduction. As division by zero is not possible, one cannot proceed. By doing row and/or column exchanges, some other non-zero element is brought to the diagonal position and made the pivot. Exchange of rows implies that the corresponding elements of $[B]$ be exchanged. This happens automatically if the augmented matrix is used. Column exchange implies that the corresponding elements in $[x]$ be exchanged. This row/column exchange to choose an appropriate pivot element is called pivoting. Even if the pivot element is non-zero, but small compared to other elements of the matrix, the accuracy of computation is reduced. So is the stability. For best results, one must do both row and column pivoting so that the largest element in the sub-matrix is chosen as the pivot.

A variation of Gaussian elimination called *LU* decomposition has several advantages and is more commonly used. This is described in the next section.

3.2 *LU* **Decomposition**

The discerning reader may already have noticed that Gaussian elimination as described above is wasteful in terms of storage requirements. At the end of Forward Reduction, the diagonal elements of $[A]$ are equal to 1 and the lower triangular elements 0. These locations could have been used to store some information. Assuming just one $(n \times n)$ matrix is used, the elements of $[A]$ (either original or modified) could have been retained in the locations where there were zeros and ones. This way the original matrix is not lost. Everytime a certain location becomes a one or a zero, the immediately preceding value is retained there. For the matrix considered in the last section, the matrix $[A]$ would look as follows at successive stages. The augmented column has been left out.

$$\begin{bmatrix} 3 & 1 & 1 \\ 1 & 1 & 1 \\ 2 & 1 & 2 \end{bmatrix} \begin{bmatrix} 3 & \dfrac{1}{3} & \dfrac{1}{3} \\ 1 & \dfrac{2}{3} & \dfrac{2}{3} \\ 2 & \dfrac{1}{3} & \dfrac{4}{3} \end{bmatrix} \begin{bmatrix} 3 & \dfrac{1}{3} & \dfrac{1}{3} \\ 1 & \dfrac{2}{3} & 1 \\ 2 & \dfrac{1}{3} & 1 \end{bmatrix}$$

It is easy to construct the original matrix $[A]$ from the reduced right most matrix given above. Denoting the elements of the reduced matrix by primed quantities

$$a_{ij} = \sum_{k=1}^{k} a'_{ik} a'_{kj} \qquad \text{if } i < j$$

$$= a'_{ij} + \sum_{k=1}^{j-1} a'_{ik} a'_{kj} \quad \text{if } i \geqslant j \qquad (3.2)$$

The above equations are arrived at just by backtracking the steps for Gaussian elimination. From these equations it also follows that

$$[A] = [L][U]$$

where

$[L] = (n \times n)$ lower triangular matrix consisting of the diagonal and lower triangular elements of the reduced matrix.

$[U] = (n \times n)$ upper triangular matrix consisting strictly of the upper triangular elements of the reduced matrix with diagonal elements equal to 1.

For the example above

$$[L] = \begin{bmatrix} 3 & 0 & 0 \\ 1 & \dfrac{2}{3} & 0 \\ 2 & \dfrac{1}{3} & 1 \end{bmatrix}$$

and

$$[U] = \begin{bmatrix} 1 & \dfrac{1}{3} & \dfrac{1}{3} \\ 0 & 1 & 1 \\ 0 & 0 & 1 \end{bmatrix}$$

We can now write a_{ij} in terms of the elements of $[L]$ and $[U]$ from Eq. [3.2]

$$a_{ij} = \sum_{k=1}^{i} l_{ik}u_{kj} \qquad \text{if } i < j$$

$$= l_{ij} + \sum_{k=1}^{j-1} l_{ik}u_{kj} \qquad \text{if } i \geqslant j$$

It is stated without proof that any non-singular matrix $[A]$ can be split into $[L]$ and $[U]$. The system of equations above can be solved to find $[L]$ and $[U]$ in terms of $[A]$. The following equations result:

$$u_{ij} = \frac{\left(a_{ij} - \sum\limits_{k=1}^{i-1} l_{ik}u_{kj}\right)}{l_{ii}} \qquad (i < j)$$

$$l_{ij} = a_{ij} - \sum_{k=1}^{j-1} l_{ik}u_{kj} \qquad (i \geqslant j) \qquad\qquad (3.4)$$

These elements have to be calculated in the proper order. The first column of $[L]$ and first row of $[U]$ are calculated first. Next, the second column and row and so on. Elements of the original matrix $[A]$ can be overwritten as elements of $[U]$ and $[L]$ are calculated. Just one array need be declared. The final matrix so obtained has $[L]$ as the lower triangular part (including the diagonal) and $[U]$ as the upper triangular part (excluding the diagonal which is equal to 1).

Once $[A]$ has been split into $[L]$ and $[U]$, solution proceeds as follows: First we solve $[L][y] = [B]$ for $[y]$ by putting $[U][x] = [y]$. Then we solve $[U][x] = [y]$ to get $[x]$. As both $[L]$ and $[U]$ are triangular, the solution is straightforward and similar to the Back Substitution step of Gaussian elimination.

LU decomposition can be done directly using Eq. (3.4). However, a more convenient algorithm for doing this on the computer is the Dolittle algorithm [1]. In order to understand how this proceeds consider a term like u_{35}

$$u_{35} = \frac{(a_{35} - l_{31}u_{15} - l_{32}u_{25})}{l_{33}}$$

As soon as the first column and row are calculated the product $l_{31}u_{15}$ is subtracted from a_{35}. This is stored in the location of a_{35}. After the second column and row are calculated the product $l_{32}u_{25}$ is subtracted from this quantity. Finally, when the third column and row are calculated the number stored in the location of a_{35} is divided by l_{33} to get u_{35}. In short, as soon as each column and row are calculated, corresponding changes are made in the remaining submatrix. These steps can be summarised as:

(1) Copy column 1 as it is (note that $l_{i1} = a_{i1}$).
(2) Divide every element of row 1, except the diagonal element, by a_{11}.
(3) For each element in the submatrix ($i > 1, j > 1$) modify a_{ij} as
$$a_{ij} = a_{ij} - a_{i1}a_{1j}$$
(4) Delete column 1 and row 1 and renumber the resulting matrix from 1. If the order is 2 or higher go to step 1. Else stop.

A Fortran segment to do the same operations would be

$$DO \quad 10 \quad K = 1, N - 1$$
$$DO \quad 20 \quad J = K + 1, N$$

$$A(K, J) = \frac{A(K, J)}{A(K, K)}$$

$$20 \quad CONTINUE$$
$$DO \quad 30 \quad I = K + 1, N$$
$$DO \quad 40 \quad J = K + 1, N$$
$$A(I, J) = A(I, J) - A(I, K)*A(K, J)$$
$$40 \quad CONTINUE$$
$$30 \quad CONTINUE$$
$$10 \quad CONTINUE$$

Based on the Dolittle algorithm; one can calculate the number of operations for LU decomposition. In calculating the first column no operations are needed. (Step 1). Step 2 involves $(n - 1)$ divisions. When considering the full $(n \times n)$ matrix, step 3 involves $(n - 1)^2$ multiplications. Total operations when considering $(n \times n)$ matrix $= (n - 1)^2 + (n - 1)$. Total operations when considering some submatrix of size $k = (k - 1)^2 + (k - 1)$. Therefore, total number of operations

$$= (n - 1)^2 + (n - 1) + (n - 2)^2 + (n - 2) + (n - 3)^2$$
$$+ (n - 3) \ldots + 2^2 + 2 + 1^2 + 1$$

$$= \sum_{i=1}^{n-1} i^2 + \sum_{i=1}^{n-1} i = \frac{(n - 1)n(2n - 1)}{6} + \frac{n(n - 1)}{2}$$

$$= \frac{n^3}{3} - \frac{n}{3}$$

Back substitution when solving for $[y]$ from $[L][y] = [B]$ takes $n(n + 1)/2$ operations as $l_{ii} \neq 1$. Solving for $[x]$ from $[U][x] = [y]$ takes $n(n - 1)/2$ operations. In all, solving a system of linear equations using LU decomposition takes

$$\frac{n^3}{3} - \frac{n}{3} + \frac{n(n + 1)}{2} + \frac{n(n - 1)}{2} = \frac{n^3}{3} + n^2 - \frac{n}{3}$$

operations. This is exactly the same as Gaussian elimination. The result is not surprising as *LU* decomposition is just a way of implementing Gaussian elimination.

However, *LU* decomposition has the following advantages:

1. If one has to solve for several different vectors [*x*] corresponding to different [*B*] but for the same [*A*], then the decomposition need be done just once. As the decomposition is the part which takes more operations this saves considerable effort.
2. The *LU* factors can be calculated in place. Elements of [*A*] can be overwritten at each step by the elements of [*L*] and [*U*]. The overwritten elements will not be needed again.
3. Properties of special matrices like real, symmetric, positive definite matrices can be better exploited [2] as explained below:

A real symmetric matrix is said to be positive definite if and only if

(a) $[x^T][A][x] > 0$ for all vectors [*x*] $\neq 0$
(b) All eigen values are positive
(c) All left half submatrices have determinants greater than zero.

It can be shown that for a real, symmetric, positive definite matrix [*A*]

$$[A] = [U^T][U] = [L][L^T]$$

i.e., it is enough if one of the factors [*L*] or [*U*] is determined. Note that unlike the earlier decomposition for a general matrix, [*U*] here has diagonal elements not necessarily equal to 1. Elements of [*U*] (or [*L*]) can be found from Cholesky's method [2].

$$u_{11} = (a_{11})^{1/2}$$

$$u_{1j} = \frac{a_{1j}}{u_{11}} \qquad\qquad j = 2 \text{ to } n$$

$$u_{ii} = \left[a_{ii} - \sum_{k=1}^{i-1} u_{ki}^2 \right]^{1/2} \qquad i = 2 \text{ to } n$$

$$u_{ij} = \frac{1}{u_{ii}} \left(a_{ij} - \sum_{k=1}^{i-1} u_{ki} u_{kj} \right) \quad \begin{array}{l} j = (i + 1) \text{ to } n \\ i = 2 \text{ to } n \end{array}$$

A DC circuit containing just linear resistors and independent sources but no controlled sources lead to real, symmetric positive definite matrices.

Pivoting is as important for *LU* decomposition as for Gaussian elimination. As in Gaussian elimination, the pivot element must be non-zero and preferably large compared to other terms in the lower right submatrix.

3.3 Exploiting Sparsity in Matrices

Whenever a network has more than about 20 nodes, the resulting matrices are almost invariably sparse. By sparse, we mean less than 30% of the elements of the matrix are non-zero. If the Tableau method were used, the matrix would be even sparser. Sparsity comes about because a node of the network is usually connected to just two or three other nodes out of a large number. Figure 3.1 gives the modified node admittance matrix (corresponding to modified node analysis) for a Biquad bandpass filter using three opamps. The upper left submatrix corresponds to node equations and voltage variables while the remaining part corresponds to constitutive relations and current variables. In all the dimension is (45×45) and only 152 out of the 2025 elements are non-zero. These are shown by 'X's. It is important that the sparsity be exploited for efficient computation. For example, operations involving zero should be avoided. Storage space can be saved by storing only the non-zero elements.

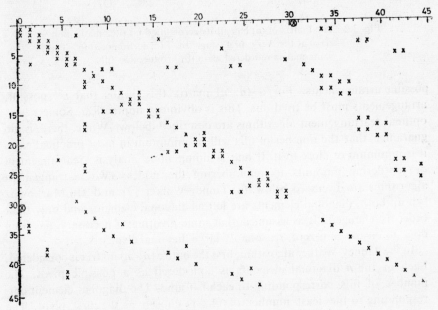

Fig. 3.1   Modified node admittance matrix for a circuit with 45 unknowns. This circuit is a Biquad band pass filter. Note that only 152 out of a total of 2025 elements are non-zero. These are shown by Xs and the zero elements not shown. The two circled entries correspond to an independent voltage source which contributes to a zero diagonal.

One problem faced in doing LU decomposition or Gaussian elimination is that even if one starts with a matrix with lots of zeros, they may get filled up. Consider the matrix of Fig. 3.2(a). At the first step of LU decomposition all the zeros will get filled up. However, rearrangement of the matrix helps in reducing the 'fills'. If the matrix of Fig. 3.2(a) is rearranged as in Fig. 3.2(b) then no 'fills' are produced. A zero element becoming a non-zero element is said to produce a fill. Once a fill is produced not only do operations involving it have to be done but the element must also be stored.

The question then arises as to how the matrix should be rearranged in order to minimise the fills. It turns out that there is no optimal strategy to minimise the number of fills. The only optimal strategy is to try out all

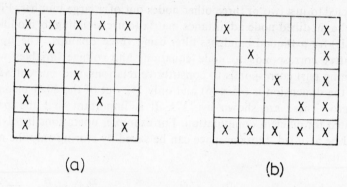

(a)   (b)

Fig. 3.2 The matrix of (a) has all its zeros filled up (i.e. made non-zero) at the very first stage of *LU* decomposition. The same matrix reordered as in (b) produces no fills.

possible arrangements. For a $(n \times n)$ matrix this means that $n^2!$ possible arrangements must be tried out. This is obviously impractical. Some nearly optimal rearrangement algorithms are described below. While there is no guarantee that the number of fills will be minimum in most practical cases it is minimum or close to it. If one is doing node analysis, rearrangement or reordering amounts to renumbering the nodes. Two rearrangement algorithms are discussed below: (a) Tinney-Walker [3] and (b) Markowitz [4]. In both, diagonal elements are left as diagonal elements and only their order is changed. It is assumed that some pivoting, if necessary, has been done to ensure non-zero, reasonably large diagonal elements.

In the Tinney-Walker algorithm, first the entire $(n \times n)$ matrix is considered. Each of the n diagonal elements is considered as a possible pivot. The number of fills corresponding to each is found. The diagonal element corresponding to the least number of fills is chosen as the first pivot. The corresponding row and column are deleted. From the remaining $((n - 1) \times (n - 1))$ matrix again the diagonal element producing the least number of fills is chosen as the pivot. The remaining $((n - 2) \times (n - 2))$ matrix is next considered and the process repeated. The algorithm can be understood from the example given below: The fills

Fills

$$
\begin{bmatrix}
a_{11} & a_{12} & 0 & a_{14} & a_{15} & 0 \\
a_{21} & a_{22} & a_{23} & 0 & a_{25} & a_{26} \\
0 & a_{32} & a_{33} & 0 & 0 & a_{36} \\
a_{41} & 0 & 0 & a_{44} & a_{45} & 0 \\
a_{51} & a_{52} & 0 & a_{54} & a_{55} & a_{56} \\
0 & a_{62} & a_{63} & 0 & a_{65} & a_{66}
\end{bmatrix}
\qquad
\begin{matrix}
a_{42}, a_{24} \\
a_{13}, a_{16}, a_{31}, a_{35}, a_{53}, a_{61} \\
\text{Nil} \\
\text{Nil} \\
a_{16}, a_{24}, a_{42}, a_{46}, a_{61}, a_{64} \\
a_{35}, a_{53}
\end{matrix}
$$

corresponding to the choice of each diagonal element as pivot are shown at the end of the row. The choice of the pivot is now limited to a_{33} or a_{44}. Where two element produce the same number of fills, the element having more non-zero elements is chosen as the pivot. This is so that the possibility of these elements causing fills further down is perhaps reduced. Both the 3rd and 4th rows (columns) contain the same number of non-zero elements. So here a_{44} is arbitrarily chosen as pivot. The fourth row and column are deleted and the process is repeated on the following submatrix:

$$
\begin{array}{cc}
\begin{array}{ccccc} 1 & 2 & 3 & 5 & 6 \end{array} & \\
\begin{array}{c} 1 \\ 2 \\ 3 \\ 5 \\ 6 \end{array}
\left[\begin{array}{ccccc}
a_{11} & a_{12} & 0 & a_{15} & 0 \\
a_{21} & a_{22} & a_{23} & a_{25} & a_{26} \\
0 & a_{32} & a_{33} & 0 & a_{36} \\
a_{51} & a_{52} & 0 & a_{55} & a_{56} \\
0 & a_{62} & a_{63} & a_{65} & a_{66}
\end{array} \right] &
\begin{array}{l}
\text{Nil} \\
a_{13}, a_{16}, a_{31}, a_{35}, a_{53}, a_{61} \\
\text{Nil} \\
a_{16}, a_{61} \\
a_{35}, a_{53}
\end{array}
\end{array}
$$

Now a_{11} is chosen as pivot. The remaining submatrix is shown below:

$$
\begin{array}{cc}
\begin{array}{cccc} 2 & 3 & 5 & 6 \end{array} & \\
\begin{array}{c} 2 \\ 3 \\ 5 \\ 6 \end{array}
\left[\begin{array}{cccc}
a_{22} & a_{23} & a_{25} & a_{26} \\
a_{32} & a_{33} & 0 & a_{36} \\
a_{52} & 0 & a_{55} & a_{56} \\
a_{62} & a_{63} & a_{65} & a_{66}
\end{array} \right] &
\begin{array}{l}
a_{35}, a_{53} \\
\text{Nil} \\
\text{Nil} \\
a_{35}, a_{53}
\end{array}
\end{array}
$$

Choosing a_{55} as pivot, we find that the remaining submatrix has no zeros. So the pivots a_{22}, a_{33}, a_{66} can be chosen in any order. The rearranged matrix looks as given below. No fills are introduced by this arrangement or ordering.

$$
\begin{array}{c}
\begin{array}{cccccc} 4 & 1 & 5 & 2 & 3 & 6 \end{array} \\
\begin{array}{c} 4 \\ 1 \\ 5 \\ 2 \\ 3 \\ 6 \end{array}
\left[\begin{array}{cccccc}
a_{44} & a_{41} & a_{45} & 0 & 0 & 0 \\
a_{14} & a_{11} & a_{14} & a_{12} & 0 & 0 \\
a_{54} & a_{51} & a_{55} & a_{52} & 0 & a_{56} \\
0 & a_{21} & a_{25} & a_{22} & a_{23} & a_{26} \\
0 & 0 & 0 & a_{32} & a_{33} & a_{36} \\
0 & 0 & a_{65} & a_{62} & a_{63} & a_{66}
\end{array} \right]
\end{array}
$$

In the Markowitz [4] algorithm we choose pivots which minimise the number of operations rather than fills. The number of operations for a_{ii} as choice of pivot from the Dolittle algorithm is

$$((\text{No. of non-zero elements in row } i) - 1)$$

((No. of non-zero elements in column i) $-$ 1)$+$
((No. of non-zero elements in row i) $-$ 1)

Let us apply the Markowitz algorithm to the matrix given above, i.e., the original matrix (not rearranged). Considering the complete $(n \times n)$ matrix we find

a_{11} pivot leads to 12 operations

a_{22} pivot leads to 20 operations

a_{33} pivot leads to 6 operations

a_{44} pivot leads to 6 operations

a_{55} pivot leads to 20 operations

a_{66} pivot leads to 12 operations.

So the choice is between a_{33} and a_{44}. Choosing a_{44} as before, we delete the 4th row and column and consider the remaining (5×5) matrix. Here the operations corresponding to a_{11}, a_{22}, a_{33}, a_{55} and a_{66} as pivots are 6, 20, 6, 12 and 12 respectively. Now we choose a_{11} or a_{33} as pivot and delete the corresponding row and column and carry on.

The example chosen to illustrate the Tinney-Walker and Markowitz algorithms was such that no fills were introduced. If a fill occurs at any stage this must be inserted and considered in the rest of the reordering procedure. Most network formulations leads to structurally symmetric matrices i.e., if $a_{ij} = 0$, then $a_{ji} = 0$. If the non-zero elements are represented by 1 and zero elements by 0, a symmetric matrix results for a structurally symmetric matrix. Of course, any matrix may be considered structurally symmetric by treating asymmetric zeros as non-zero elements. But the sparsity is not then fully exploited. For a structurally symmetric matrix, the Markowitz algorithm just amounts to choosing that diagonal element as pivot which has the least number of off-diagonal non-zero elements in its corresponding row and column. This is done first for the $(n \times n)$ matrix, next for the remaining $((n - 1) \times (n - 1))$ matrix and so on.

In the rearrangement discussed above, it has been assumed that the matrix has a strong diagonal. This is true for node analysis, but may not be true for other formulations. The matrix shown in Fig. 3.1 resulted from modified node analysis. Independent voltage sources with one terminal grounded give rise to a row and a column with just one symmetric non-zero entry each and with the common diagonal element zero. Figure 3.1 shows these non-zero elements circled for one voltage source. The rows (or columns) corresponding to these two entries must be exchanged to get rid of the zero diagonal.

Floating independent voltage sources must also have their branch constitutive relation exchanged with the node equation at one of the two nodes. For more details on modified node analysis one is referred to [5]. In general, whatever be the formulation technique, some initial rearrangement must be done to get a strong diagonal. Then Tinney-Walker, Markowitz or any other technique may be applied to reduce fills.

In solving sparse matrices an efficient data structure to store the non-zero elements is as important as reordering to minimise fills. As a first step, no longer one uses a two-dimensional array but an one-dimensional array. An one-dimensional array with only the non-zero elements stored is not by itself sufficient. Two additional integer arrays would be needed to store the row and column number. Thus, for each non-zero element the row number, column number and value would be stored. This data structure, though simple, is not very convenient either for reordering or for *LU* decomposition. Linked lists with suitable pointers would be far more convenient. For example, if a fill occurred a large number of elements would have to be shifted to insert the fill. Other operations like searching a column or a row for a non-zero element or deleting a row or column are also not easily done. Two possible data structures to facilitate the non-numeric operations involved in the Tinney-Walker or Markowitz algorithm are given below.

In the first, the off-diagonal non-zero elements are numbered in some arbitrary order. The following arrays are created:

ROW(I)—row number of ith element

COL(I)—column number of ith element

BR (I)—number of non-zero element in the same row as ith element but just before it

AR(I)—number of non-zero element in the same row as ith element but just after it

BC(I)—number of non-zero element in the same column as ith element but just before it

AC(I)—number of non-zero element in the same column as ith element but just after it.

The arrays $BR(I)$, $AR(I)$, $BC(I)$ and $BR(I)$ define the non-zero elements surrounding the ith element. For a structurally symmetric matrix, BR and AR may be sufficient. Alternatively, these arrays need be constructed for just half the number of total off diagonal non-zero elements. The diagonal elements are not involved in the reordering and have therefore been left out in the above scheme for storage. Of course, they will be needed when the actual *LU* decomposition is done.

As an example, consider the matrix shown in Fig. 3.3(a). Various non-zero elements have been *numbered* in some arbitrary order as shown in Fig. 3.3(b). The array entries corresponding to element number 6 are

ROW(6) = 2

COL(6) = 3

BR(6) = 4

AR(6) = 0 (no element on right in the same row)

BC(6) = 2

AC(6) = 0 (no element below in the same column).

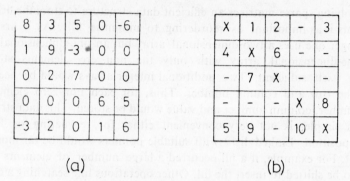

Fig. 3.3 In (a) a matrix is shown. In (b) the numbering of the off-diagonal non-zero elements is given. In this scheme diagonal elements are not numbered.

A second scheme for the storage of non-zero elements with suitable pointers is given in reference [6]. The discussion given here is for structurally symmetric matrices. The following arrays are used.

DIAG contains the diagonal elements stored in the order $a_{11}, a_{22} \ldots a_{nn}$, UPACK contains off-diagonal non-zero elements above the diagonal in row order. LPACK contains off-diagonal, non-zero elements below the diagonal in column order. UCOL gives the column number of elements in UPACK, UROWST gives the element number in UPACK from which each row starts.

Fills must be anticipated in this storage scheme unlike the other scheme discussed above. Array entries must be made for the fills with the value 0. This is because insertion is more tedious here with the elements numbered sequentially in row or column order. Consider the matrix shown in Fig. 3.3(a). By assuming a_{31} and a_{25} to be nonzero, structural symmetry is obtained. Let us assume that a_{35} and a_{53} are going to be 'filled'. These elements are also therefore stored. The various arrays then are as follows: The entry UROWST(5) needs some explanation. Its value is given as 8 though UPACK has only 7 elements. For the last row as well as rows with no entry in UPACK some definition must be made. One possibility is to give the value 0 for UROWST of the corresponding row. Here value 8 indicates that no element from row 5 exists in UPACK and that it is possibly the last row.

DIAG (1) = 8	UPACK (1) = 3	LPACK (1) = 1
DIAG (2) = 9	UPACK (2) = 5	LPACK (2) = 0
DIAG (3) = 7	UPACK (3) = −6	LPACK (3) = −3
DIAG (4) = 6	UPACK (4) = −3	LPACK (4) = 2
DIAG (5) = 6	UPACK (5) = 0	LPACK (5) = 2
	UPACK (6) = 0	LPACK (6) = 0
	UPACK (7) = 5	LPACK (7) = 1

UCOL (1) = 2	UROWST (1) = 1
UCOL (2) = 3	UROWST (2) = 4
UCOL (3) = 5	UROWST (3) = 6
UCOL (4) = 3	UROWST (4) = 7
UCOL (5) = 5	UROWST (5) = 8
UCOL (6) = 5	
UCOL (7) = 5	

In short, sparsity is exploited in two ways. First, by reordering to mini-mise fills and secondly by evolving suitable data structures for efficient and convenient storage of the non-zero elements. The operations involved in reordering are strictly non-numeric and should therefore take far less computer time. Before reordering, pivoting must be done to ensure a strong diagonal.

3.4 Summary

Solving any network finally boils down to the solution of a set of simul-taneous linear algebraic equations. The most common method used to solve these equations is *LU* decomposition which is just a variation of Gaussian elimination. Whatever be the formulation used for the equations, the resulting equations will usually be sparse i.e. the associated matrix will have a large number of zeros. This sparsity must be exploited in solving the equations. This is done by first evolving a suitable data structure and storing only the non-zero elements. Then the matrix elements are rearranged so that fills are avoided as far as possible during the solution process. A fill is a conversion of a zero element into a non-zero element. The rearrange-ment involves non-numeric operations and should therefore be fast. At all times care must be taken to ensure that diagonal elements are non-zero and relatively large.

References

1. L.O. Chua and P. Lin, Computer-Aided Analysis of Electronic Circuits, p. 638, Prentice-Hall, Inc., 1975.
2. J.R. Westlake, A. Handbook of Numerical Matrix Inversion and Solution of Linear Equations, John Wiley and Sons, 1968.
3. W.F. Tinney and J.W. Walker, Direct Solutions of Sparse Network Equations by Optimally Ordered Triangular Factorisation, Proc. IEEE, vol. 55, pp. 1801-1809, Nov. 1967.
4. H.M. Markowitz, The Elimination Form of the Inverse and its Application to Linear Programming, Management Science, vol. 3, pp. 255-269, 1957.
5. C.W. Ho, A.E. Ruehli and P.A. Brennan, The Modified Nodal Approach to Net-work Analysis, IEEE Trans. Ccts. and Syst , vol. CAS-22, pp. 504-509.
6. R.D. Berry, An Optimum Ordering of Electronic Circuit Equations for a Sparse Matrix Solution, IEEE Trans. Circuit Theory, vol. CT-18, pp. 40-50, Jan. 1971.

Problems

1. For a DC network having only linear resistors and independent sources, show that the node admittance matrix is real, symmetric and positive definite.

2. The Gauss-Jordan elimination method is similar to Gaussian elimination except that it reduces the given matrix $[A]$ to a diagonal one with all diagonal elements $= 1$. Calculate the number of operations required to solve a set of simultaneous linear equations using this method.

3. Assume that the inverse of $[A]$ is available. Show that for every new $[B]$, n^2 operations are required to obtain the solution from $[A^{-1}][B]$.

4. Show that if a matrix $[A]$ is a product of two matrices $[L]$ and $[U]$ with

$[L] =$ lower triangular

$[U] =$ upper triangular with all diagonal elements $= 1$

then

$$a_{ij} = \sum_{k=1}^{i} l_{ik} u_{kj} \qquad \text{for } i < j$$

$$= l_{ij} + \sum_{k=1}^{j-1} l_{ik} u_{kj} \qquad \text{for } i \geqslant j$$

5. Reorder the matrices given below to minimise the number of fills using (a) Tinney-Walker algorithm (b) Markowitz algorithm

```
x   x   0   x   0            x   0   0   x   x   0
x   x   x   0   x            0   x   x   x   0   0
0   x   x   0   0            x   x   x   0   x   0
x   0   0   x   0            x   0   0   x   0   x
0   x   0   0   x            0   0   x   0   x   0
                            0   0   x   x   0   x
```

```
x   0   0   0   0   0
x   x   0   x   0   0
0   0   x   x   0   0
0   x   x   x   0   0
0   x   x   0   x   0
0   0   0   x   0   x
```

In each case identify the fills which would occur if the equations were solved without reordering.

6. Find the LU factors for the two matrices given below using the Dolittle algorithm. Do pivoting if necessary.

```
 1   0   0  -2            0   1   0
 0   2   7   0           -1   3  -1
-3   6   3   0            2  -3   2
 5   0   0   4
```

7. Show that reordering for Gaussian elimination and LU decomposition can be done using the same techniques.

8. In doing modified node analysis formulation, many of the branch constitutive relations lead to zero diagonal elements. Pivoting must therefore be done before reordering and LU decomposition. Evolve rules for pivoting corresponding to the branch constitutive relations of the following elements: (a) independent voltage

source (b) CCVS (c) VCVS. The elements may have one terminal grounded or neither terminal grounded. Consider both cases.

9. Represent the matrices of Problems 5 and 6 using the data structure of the second scheme given in Section 3.3.

Programming Assignments

1. A program based on modified node analysis for linear circuits was given at the end of Chapter 2. For this program write your own routine for solving linear equations rather than use the IMSL routine LEQT1F. First do pivoting to take care of the zero diagonal elements. These will occur in the rows corresponding to the branch constitutive relations for voltage sources. Then reorder the matrix to minimise fills. Finally, apply the Dotittle algorithm for *LU* decomposition. You should use a sparse matrix data structure for storage when you do the reordering.

2. Compare the time and storage taken in programming assignment 1 for your linear equation solver for sparse matrices as opposed to LEGT1F which is not meant for sparse matrices.

4

DC Analysis of Non-linear Circuits

Virtually any electronic circuit is non-linear. Relatively simple semiconductor elements like diodes are highly non-linear. The techniques developed in Chapters 2 and 3 are, therefore, not of much use till they can be extended to non-linear circuits. This is done in this chapter; but the discussion is restricted to two terminal non-linear elements. Any multi-terminal non-linear device like a transistor would have to be represented in terms of two terminal elements like diodes. Both non-linear resistors and controlled sources are considered. As the chapter is on DC analysis, capacitors and inductors are not of interest. Initially the treatment is based on node analysis. Therefore, non-linear resistors and non-linear VCCS' alone are considered. Later on, the treatment is extended to all four kinds of non-linear controlled sources. Node analysis restricts one to voltage controlled non-linear elements because voltages are the unknowns.

Any DC analysis of a non-linear circuit leads to a set of simultaneous non-linear algebraic equations. These usually have to be solved by an iterative technique after an initial guess. Some criterion must be used to test for convergence. The solution technique must ensure convergence in most cases.

4.1 Non-Linear Equation in One Unknown

Figure 4.1 shows a circuit with one non-linear resistor, a diode. The only unknown to be solved for is V_1. The $i-v$ relation for the diode is given by

$$I_d = I_s \left(\exp \left(\frac{V_d}{V_T} \right) - 1 \right)$$

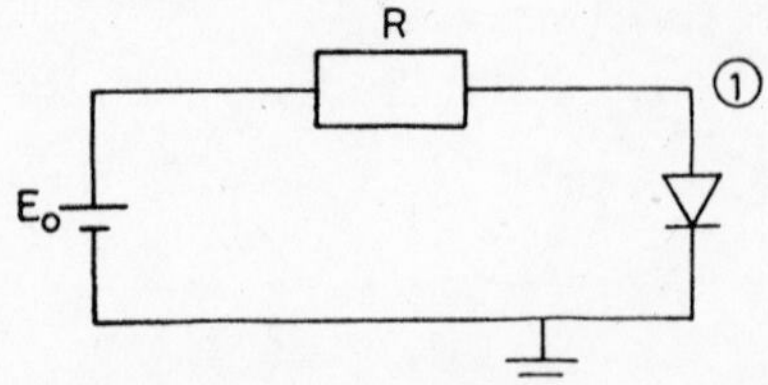

Fig. 4.1 Simple circuit with one non-linear element—the diode.

where

I_s = reverse saturation current

$V_T = kT/q$ which is approximately 26 mV at room temperature.

As defined above, the diode is a voltage controlled non-linear resistor. One can find V_1 by writing the node equation at node 1

$$\frac{V_1 - E_0}{R} + I_s\left(\exp\left(\frac{V_1}{V_T}\right) - 1\right) = 0$$

This is of the form $f(V_1) = 0$. A non-linear equation $f(x) = 0$ can be solved by a number of techniques. Among these are [1]

1. Newton-Raphson (N-R)
2. Secant method
3. Fixed Point Iteration
4. Regula Falsi

All the methods involve making an initial guess and doing iterations. In fact, the secant method requires two guesses. Of the methods listed above, the N - R technique alone is applicable to a system of equations. It also has good convergence properties and is almost always used to solve a system of non-linear equations. For a single unknown, it can be given a geometric interpretation as shown in Fig. 4.2.

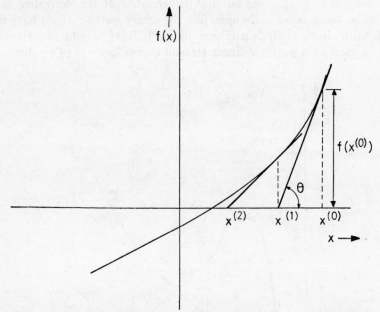

Fig. 4.2 Graphical interpretation of the Newton-Raphson Technique.

First, an initial guess $x^{(0)}$ is made. The function $f(x)$ is expanded around $x^{(0)}$ using the Taylor series. Taking just the first term of the series

$$f(x) = f(x^{(0)}) + f'(x)(x - x^{(0)})$$

where $f'(x)$ is evaluated at $x = x^{(0)}$.

Solving the above equation by equating to zero gives the first iterate. Solution is easy as the equation is linear and one gets

$$x^{(1)} = x^{(0)} - \frac{f(x^{(0)})}{f'(x)_{/x=x^{(0)}}}$$

After the jth iteration, one gets

$$x^{(j)} = x^{(j-1)} - \frac{f(x^{(j-1)})}{f'(x)_{/x=x^{(j-1)}}}$$

Figure 4.2 shows geometrically how $x^{(1)}$ is obtained from $x^{(0)}$ and how one converges to the solution.

There is no guarantee that the solution always converges. One obvious situation where the solution fails to converge is given in Fig. 4.3. If the initial guess is made in the valley, one cannot approach the solution. Often, more than one solution exists. In such situations different initial guesses lead to different solutions. Most of the time, fortunately, just one solution exists for a given network. Circuits, like flip-flops, which have two or more solutions might need a suitable initial condition or initial guess to be specified. The Newton-Raphson technique not only needs the function $f(x)$ to be specified but also its derivative. If the derivative is not available, some other technique like the secant method would have to be used. Most circuit analysis packages, like SPICE [1], would not allow the user to specify his own non-linear element except in terms of a polynomial.

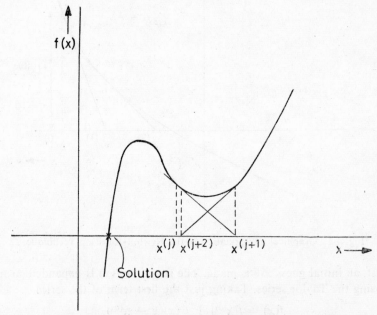

Fig. 4.3   A situation where Newton-Raphson fails to converge.

This is because differentiation of an arbitrary function is a difficult exercise in symbol manipulation in a computer. Common elements like diodes, BJTs and MOSFETs have built in models along with the necessary derivative functions in SPICE [1].

One can give a circuit interpretation to the N-R technique which makes it very convenient for solving circuits. Applying N-R to the circuit of Fig. 4.1, one gets for the jth iteration

$$V_1^{(j)} = V_1^{(j-1)} - \frac{f(V_1^{(j-1)})}{f'(V_1)_{/V1=V1^{(j-1)}}}$$

where

$$f(V_1^{(j-1)}) = \frac{V_1^{(j-1)} - E_0}{R} + I_S\left(\exp\left(\frac{V_1^{(j-1)}}{V_T}\right) - 1\right)$$

and

$$f'(V_1) = \frac{1}{R} + \frac{I_S}{V_T}\left(\exp\left(\frac{V_1}{V_T}\right)\right)$$

On rewriting, we get

$$\frac{V_1^{(j)} - E_0}{R} = V_1^{(j-1)} \frac{I_S}{V_T}\left(\exp\left(\frac{(V_1^{(j-1)})}{V_T}\right)\right) - V_1^{(j)}\frac{I_S}{V_T}\left(\exp\left(\frac{(V_1^{(j-1)})}{V_T}\right)\right)$$

$$-I_S\left(\exp\left(\frac{V_1^{(j-1)}}{V_T}\right) - 1\right) \tag{4.1}$$

Let

$$\frac{I_S}{V_T}\left(\exp\left(\frac{(V_1^{(j-1)})}{V_T}\right)\right) = G^{(j-1)}$$

and

$$I_S\left(\exp\left(\frac{V_1^{(j-1)}}{V_T}\right) - 1\right) - V_1^{(j-1)}\frac{I_S}{V_T}\left(\exp\left(\frac{(V_1^{(j-1)})}{V_T}\right)\right) = I^{(j-1)}$$

Then Eq. (4.1) above represents the linear network shown in Fig. (4.4) where $V_1^{(j)}$ is to be solved for. The N-R technique, at least for this

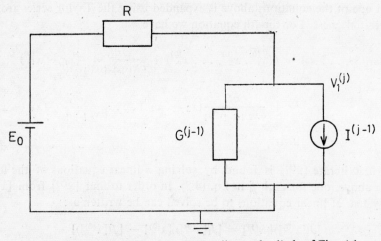

Fig. 4.4 Linearised equivalent corresponding to the diode of Fig. 4.1.

example, amounts to replacing the non-linear resistor by a linear conductor $G^{(j-1)}$ in parallel with an independent current source $I^{(j-1)}$ at each iteration. The value of the linear resistor (or conductor) is the incremental value at the voltage from the previous iteration $V_1^{(j-1)}$. The value of the independent current source is given by

$$I^{(j-1)} = I_d - G^{(j-1)}V_1^{(j-1)}$$

where I_d is evaluated at $V_d = V_1^{(j-1)}$ and is the current through the diode for some V_d.

The linear replacement for the non-linear element at each iteration is called the linearised equivalent [3]. This concept is extended in Section 4.3 to a non-linear network with many nodes leading to a system of non-linear algebraic equations.

4.2 N-R Techniques for Many Variables

Consider a system of n non-linear equations in n unknowns

$$f_1(x_1, x_2, \ldots x_n) = 0$$

$$f_2(x_1, x_2, \ldots x_n) = 0$$

$$\cdot$$
$$\cdot$$
$$\cdot$$

$$f_n(x_1, x_2, \ldots x_n) = 0 \qquad (4.2)$$

Let an initial guess be made for the vector

$$[x] = [x^{(0)}] = [x_1^{(0)}, x_2^{(0)} \ldots x_n^{(0)}].$$

Each one of the equations above is expanded using the Taylor series around this initial guess. For the ith equation we have

$$f_i(x_1, x_2, \ldots x_n) = f_i(x_1^{(0)}, x_2^{(0)}, \ldots x_n^{(0)}) + \frac{\partial f_i}{\partial x_{1/[x]=[x^{(0)}]}}(x_1 - x_1^{(0)})$$

$$+ \frac{\partial f_i}{\partial x_{2/[x]=[x^{(0)}]}}(x_2 - x_2^{(0)}) \ldots \frac{\partial f_i}{\partial x_{n/[x]=[x^{(0)}]}}(x_n - x_n^{(0)})$$

$$= 0$$

The first iterate $[x^{(1)}]$ is found by solving n linear equations of the form given above, one for each f_i in Eq. (4.2). In order to find $[x^{(1)}]$ from $[x^{(0)}]$, the system of linear equations to be solved can be written as:

$$[J([x^{(0)}])][x^{(1)}] = [J([x^{(0)}])][x^{(0)}] - [f([x^{(0)}])]$$

where $J([x^{(0)}])$ is called the Jacobian and given by

$$\begin{bmatrix} \dfrac{\partial f_1}{\partial x_1} & \dfrac{\partial f_1}{\partial x_2} & \cdots & \dfrac{\partial f_1}{\partial x_n} \\[2ex] \dfrac{\partial f_2}{\partial x_1} & \dfrac{\partial f_2}{\partial x_2} & \cdots & \dfrac{\partial f_2}{\partial x_n} \\[2ex] \vdots & & & \\[2ex] \dfrac{\partial f_n}{\partial x_1} & \dfrac{\partial f_n}{\partial x_2} & \cdots & \dfrac{\partial f_n}{\partial x_n} \end{bmatrix}_{/[x]=[x^{(0)}]}$$

The iterate $[x^{(j)}]$ is found from $[x^{(j-1)}]$ using the equation

$$[J([x^{(j-1)}])][x^{(j)}] = [J([x^{(j-1)}])][x^{(j-1)}] - [f([x^{(j-1)}])]$$

Iterations are continued till the solution converges i.e. two successive iterations are close as defined by some criterion.

We see that the N-R technique for solving a system of n non-linear equations amounts to solving a system of n linear equations many times iteratively till convergence is reached. At each iteration n^2 derivatives have to be calculated. This appears to be a formidable task. Fortunately, in solving non-linear equations resulting from non-linear circuits, most of the derivatives are zero or constant. It was pointed out in Chapter 3, for a large circuit, a node is connected only to a few other nodes. In node analysis each of the f_i would represent a node equation and would only be a function of a small number of the total number of node voltages. Also each of the VCCS', in node analysis, would normally be a function of just one controlling branch. Each f_i would involve terms which are functions of unknown branch voltages of the form $f(V_k - V_l)$. Here V_k and V_l represent node voltages across a branch. In short, most of the n^2 derivatives need not be calculated. This should be exploited in solving a network using the N-R technique. The linearised equivalent is a convenient way of doing this and is the subject of discussion in the following section:

4.3 Linearised Equivalent Circuit for the N-R Technique

The discussion in this section is restricted to node analysis. Voltage controlled non-linear resistors and VCCS' are the only non-linear elements allowed. The next section extends the same methods to a hybrid formulation techniques like modified node analysis so that other kinds of non-linear elements can be handled.

Consider a node n connected to node 1 by a linear resistor R, to node 4 by an independent current source I_s, to node 3 by a non-linear VCCS $g_m(V_c)$ and to node 2 by a non-linear resistor $g(V)$ (i.e., $I = g(V)$ for the non-linear resistor). Figure 4.5(a) shows the circuit segment. The non-linear node equation at node n gives

$$\frac{V_n - V_1}{R} + g + g_m - I_s = 0 = f_n$$

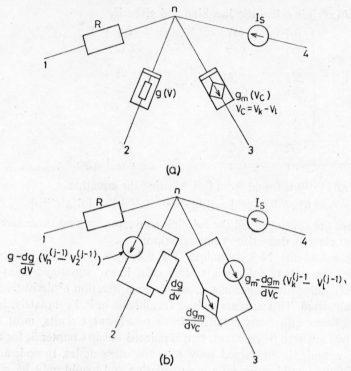

Fig. 4.5 (a) A general node n with various elements connected to it.
(b) Linearised equivalent corresponding to (a).

where g is evaluated at $V = (V_n - V_2)$ and g_m at $V_c = (V_k - V_l)$.
$(V_k - V_l)$ is the voltage across the controlling branch. As discussed in the last section, it is clear that the non-linear terms are functions of a branch voltage i.e. a function of the difference of two node voltages. For the function f_n above, we can write

$$\frac{\partial f_n}{\partial V_n} = \frac{\partial g}{\partial V_{/V=(Vn-V2)}} + \frac{1}{R} = \frac{dg}{dV_{/V=(Vn-V2)}} + \frac{1}{R}$$

$$\frac{\partial f_n}{\partial V_2} = \frac{-\partial g}{\partial V_{/V=(Vn-V2)}}$$

$$\frac{\partial f_n}{\partial V_k} = \frac{\partial g_m}{\partial V_{c/Vc=(Vk-Vl)}} = \frac{dg_m}{dV_{c/Vc=(Vk-Vl)}}$$

$$\frac{\partial f_n}{\partial V_l} = \frac{-\partial g}{\partial V_{c/Vc=(Vk-Vl)}}$$

Chain rules [4] have been used in deriving the above equations. In order to find $\dfrac{\partial f_n}{\partial V_n}$, $\dfrac{\partial g}{\partial V_n}$ must be calculated. First g is considered a function of V and V is then taken as a function of two variables V_n and V_2. Values for the jth iteration are found by expanding the function f_n using the Taylor series and values from the $(j-1)$th iteration.

$$\frac{V_n^{(j-1)} - V_1^{(j-1)}}{R} + g + g_m - I_s + (V_n^{(j)} - V_n^{(j-1)})\left[\frac{1}{R} + \frac{dg}{dV}\right]$$

$$-(V_2^{(j)} - V_2^{(j-1)})\frac{dg}{dV} - \frac{(V_1^{(j)} - V_1^{(j-1)})}{R}$$

$$+ \frac{dg_m}{dV_c}[(V_k^{(j)} - V_k^{(j-1)}) - (V_l^{(j)} - V_l^{(j-1)})] = 0$$

Rewriting, we get

$$\frac{(V_n^{(j)} - V_1^{(j)})}{R} + \frac{dg}{dV}(V_n^{(j)} - V_2^{(j)}) + \frac{dg_m}{dV_c}(V_k^{(j)} - V_l^{(j)}) - I_s + g$$

$$+ g_m - \frac{dg}{dV}(V_n^{(j-1)} - V_2^{(j-1)}) - \frac{dg_m}{dV_c}(V_k^{(j-1)} - V_l^{(j-1)}) = 0$$

This equation corresponds to the circuit shown in Fig. 4.5(b). One can, therefore, give a circuit interpretation to each $N - R$ iteration as follows. For each iteration, the original non-linear network is replaced with the linearised equivalent according to the following rules.

1. Replace every non-linear resistor ($I = g(V)$) by a linear resistor whose conductance G for the jth iteration is

$$G^{(j-1)} = \frac{dg}{dV_{/V=V^{(j-1)}}}$$

where $V^{(j-1)}$ is the branch voltage after the $(j-1)$th iteration.

2. Add an independent current source in parallel to $G^{(j-1)}$ given by

$$I^{(j-1)} = g - G^{(j-1)}V^{(j-1)}.$$

If $V = (V_{n1} - V_{n2})$, where $n1$ and $n2$ are nodes between which the non-linear resistor is connected then $I^{(j-1)}$ flows from $n1$ to $n2$. In the expression for $I^{(j-1)}$ above, g is evaluated at $V^{(j-1)}$.

3. Similarly replace every non-linear VCCS ($I = g_m(V_c)$) by a linear VCCS $G_m^{(j-1)}$ given by

$$G_m^{(j-1)} = \frac{dg_m}{dV_{c/Vc=Vc^{(j-1)}}}$$

where $V_c^{(j-1)}$ is the controlling branch voltage after the $(j-1)$th iteration.

4. Add a current source in parallel to $G_m^{(j-1)}$ given by

$$I_m^{(j-1)} = g_m - G_m^{(j-1)} V_c^{(j-1)}$$

Example 4.1:   The transistor circuit shown in Fig. 4.6(a) is to be analysed. The transistor is replaced by a non-linear resistor between base and emitter and a non-linear VCCS between base and collector. This is just the Ebers-Moll model simplified for the normal active mode. The non-linear circuit in Fig. 4.6(b) is to be solved for V_B and V_C. For simplicity, let the diode non-linearity $g(V)$ be given by the piecewise linear model of Fig. 4.6(c).

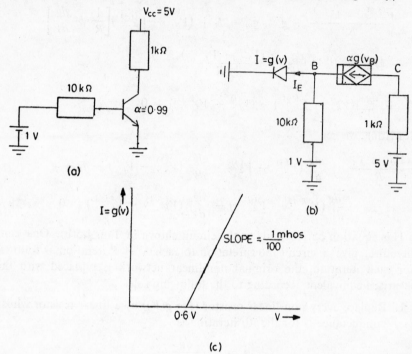

Fig. 4.6 (a) Simple transistor circuit; (b) Model for circuit in (a); (c) Piecewise linear model for the diode.

The N–R technique requires an initial guess. Let this be $V_B^{(0)} = 0$ V and $V_C^{(0)} = 5$V. The linear circuit to be solved for to get $V_B^{(1)}$, $V_C^{(1)}$ is shown in Fig. 4.7(a). Solving we get $V_B^{(1)} = 1$V, $V_C^{(1)} = 5$V. Fig. 4.7(b) is the new linearised equivalent. Writing node equations at B and C, we get

$$\frac{1 - V_B^{(2)}}{10K} = \left[4mA - 10mA + \frac{V_B^{(2)}}{0.1k} \right] - \frac{0.99 V_B^{(2)}}{0.1k}$$

$$-0.99 \times 4mA + \frac{0.99}{0.1k}$$

$$\frac{V_C^{(2)} - 5}{1k} + \frac{0.99}{0.1k} V_B^{(2)} + 0.99 \times 4mA - \frac{0.99}{0.1k} = 0$$

On solving it we get $V_B^{(2)} = 0.8$ V, $V_C^{(2)} = 0.3$ V. This is the exact solution. This was possible in just two iterations because of the simple piecewise linear diode model. In general, the solution is never exact though one may get arbitrarily close by doing more iterations.

4.4 Linearised Equivalent Circuit for Hybrid Formulations

If node analysis is used, voltage controlled non-linear elements alone can

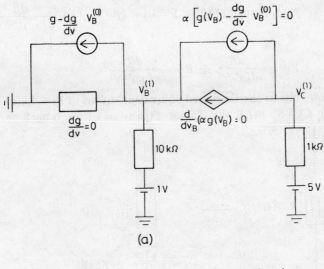

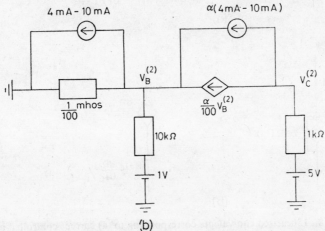

Fig. 4.7 (a) Linearised equivalent for the first iteration corresponding to Fig. 4.6. (b) Linearised equivalent for second iteraction.

be allowed. Further, of the four kinds of controlled sources (linear or non-linear) VCCS' alone can be analysed. Let us therefore use a hybrid formulation, namely modified node analysis (MNA), described in Chapter 2.2 for linear circuits. It is simply extended to non-linear circuits by adding that any non-linear element (resistor or controlled source) controlled by a current must have that current chosen as an unknown. This is really implied by the rules stated in Section 2.2.3.

Linearised equivalents for these various additional non-linear elements are easily obtained. Figure 4.8 shows the linearised equivalents for a current controlled non-linear resistor and for a non-linear CCVS. For the non-linear resistor, the non-linear function $(V - R(I) = 0)$ will occur in the branch constitutive relation. Expanding this in a Taylor series around $(I^{(j-1)}, V^{(j-1)})$ we get

$$V^{(j-1)} - R_{/I=I}^{(j-1)} + (V^{(j)} - V^{(j-1)}) - (I^{(j)} - I^{(j-1)})\frac{dR}{dI_{/I=I}^{(j-1)}} = 0$$

or

$$V^{(j)} = \frac{dR}{dI_{/I=I}^{(j)}} + R - \frac{dR}{dI_{/I=I}^{(j-1)}}$$

This corresponds to the linearised equivalent shown in Fig. 4.8(a). For the non-linear CCVS, $R(I_c)$, the Taylor expansion of the branch constitutive

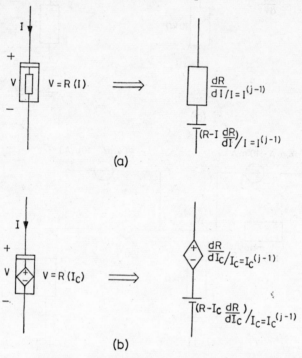

(a)

(b)

Fig. 4.8   Linearised equivalents corresponding to (a) current controlled non-linear resistor (b) non-linear CCVS.

relation would give

$$V^{(j)} = R - \frac{dR}{dI_{c/Ic=Ic}^{(j-1)}} + \frac{dR}{dI_{c/Ic=Ic}^{(j)}}$$

where R and $\frac{dR}{dI_c}$ are evaluated at $I_c^{(j-1)}$.

This corresponds to a linear CCVS in series with an independent voltage source as shown in Fig. 4.8(b). Similarly linearised equivalents can be found for non-linear VCVS' and CCCS'.

Example 4.2:   Example 4.1 can be reworked using MNA. Figure 4.9(b) shows the circuit with the simplified Ebers-Moll equivalent circuit for the transistor incorporated. Now we have a linear CCCS between B and C. Let the diode be represented by the piecewise linear model of Fig. 4.9(c). It will be considered as a current controlled non-linear element now.

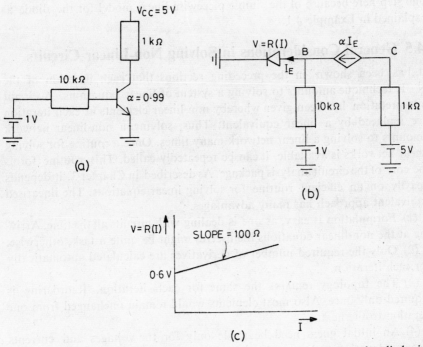

(a)

(b)

(c)

Fig. 4.9 The same circuit as that in Fig. 4.6 is analysed considering the diode in (b) as a current controlled non-linear element. The axes in (c) are therefore interchanged as compared with Fig. 4.6(c).

Let the initial guess for $I_E^{(0)}$ be 10 mA. The diode is replaced by the linearised equivalent shown in Fig. 4.10. The network of Fig. 4.10 can be solved using MNA. On solving it we get $I_E^{(1)} = 2$ mA. Using this value as

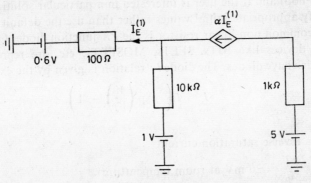

Fig. 4.10 Linearised equivalent for first iteration corresponding to Fig. 4.9. The diode is considered to be a current controlled non-linear element.

the initial guess a new linearised equivalent can be constructed. But this will lead to the same circuit and the same answer. Convergence takes place in one step here because of the simple piecewise linear model for the diode as explained in Example 4.1.

4.5 General Considerations in Solving Non-Linear Circuits

It has been shown in the preceding sections that each iteration of the N – R technique amounts to solving a system of linear equations. A circuit interpretation has been given whereby non-linear elements at each iteration are replaced by a linear equivalent. Thus, solving a non-linear network amounts to solving a linear network many times. Once a routine for solving linear networks is available, it can be repeatedly called. This routine forms the core of the circuit analysis package. As described in Chapter 3, it depends heavily on an efficient routine for solving linear equations. The linearised equivalent approach has many advantages.

(1) Formulation is easy, as one is dealing with circuits all the time. Arriving at the non-linear equations themselves might be quite a task, otherwise.

(2) Only the required number of derivatives are calculated automatically for each iteration.

(3) The topology remains the same for each iteration. Reordering is required only once. Also most elements would remain unchanged from one iteration to the next.

(4) An initial guess need be made only for the voltages and currents associated with non-linear elements. In example 4.2 an initial guess was made for I_E alone.

One may wonder how initial guesses may be given in a general circuit analysis program. It would involve more work on the part of the user, if he had to specify them. He may even be unaware of the techniques used. Often the program itself chooses initial values. A convenient guess is zero and it works most of the time. If the circuit has more than one solution (like a flip-flop) and if the user is interested in a particular solution then he may specify appropriate initial values rather than use the default ones.

A very common non-linear resistor is the PN junction diode. Other semiconductor devices like BJTs, JFETs, MOSFETs, etc. are represented by models which have diodes. The diode i-v relation is given by the exponential.

$$I = I_s\left(\ \exp\left(\frac{V}{V_T}\right) - 1\right)$$

where

I_s = reverse saturation current

$V_T = \dfrac{kT}{q} \simeq 26$ mV at room temperature

Unfortunately, this is a very awkward function to handle leading to overflows and underflows on a computer. Consider the simple circuit of Fig. 4.1. Let the initial guess for $V_1(= V_1^{(0)})$ be 0 V. The diode is to be replaced by a conductance $\dfrac{dI}{dV}_{|V=0}$ in parallel with an independent current source $I^{(0)}$.

For $V_1^{(0)} = 0$ V, $I^{(0)} = 0$ and $dI/dV = I_S/V_T$. Usually I_S is very small, of the order of 10^{-14} A. So the conductance value is very small. Solving one gets $V_1^{(1)} \simeq E_0$.

Now let $E_0 = 5$ V. For the second iteration the quantity $\exp\left(\dfrac{V_1^{(1)}}{V_T}\right)$ must be calculated. But this is very large and will cause an overflow in a computer. Similarly, if the voltage across the diode is a large negative value (like -5 V) then an underflow will occur. Two possible ways of overcoming these are given below:

(1) The exponential relation for a diode is an ideal relation and is usually only valid for small forward and reverse bias voltages. For large forward bias the current will be restricted by the contact and bulk resistance. For reverse bias, the current will be greater than what the exponential relation predicts because of surface leakage. An actual diode may be more accurately represented by a model like that in Fig. 4.11(a). A way of incorporating this model is to redefine the i-v relations as shown in Fig. 4.11(b). Between A and B the exponential relation is retained. Beyond A a straight line characteristic is used with the slope equal to that of the exponential at A. To the left of B, another straight line characteristic is used, again with a slope equal to that of the exponential at B. Points A and B are so chosen that $\exp(V_A/V_T)$ and $\exp(V_B/V_T)$ can be represented comfortably on a computer without overflows and underflows. Usually the solution, for forward bias, would converge between O and A and the representation to the right of A is immaterial.

(2) Alternatively, the diode can be considered as a current controlled non-linear element with

$$V = V_T \ln\left[\frac{I}{I_s} + 1\right]$$

This works fine as long as I does not take a negative value greater than I_s. Then one faces the problem that the logarithm of a negative number is not defined. Another disadvantage of this approach is that convergence is a lot slower than for the exponential function (see Problem 4.8). Also, node analysis can no longer be used.

Convergence is critically related to the choice of functions used to describe the non-linearity in a device. A convenient approach is to use a piecewise linear model to describe the non-linearity. This was done for the diode in Examples 4.1 and 4.2. By increasing the number of linear segments, the i-v characteristics can be represented more accurately. Derivatives are easy to calculate and their values can be kept within bounds. Once one reaches the segment where the final solution lies, convergence takes place in one step. As mentioned in Section 4.1, a user defined non-linearity must have a form conveniently differentiable. A piecewise linear model is one such form. So is a polynomial representation.

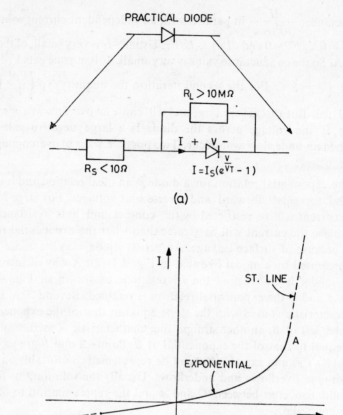

Fig. 4.11   (a) Model for a practical diode in terms of an ideal diode.
(b) Model for the *i-v* characteristics of a diode to avoid
overflows and underflows in a computer.

In this chapter, so far, we have been considering non-linear controlled sources which have just one branch voltage or current as a controlling quantity. Often there may be more than one controlling quantity. Linearised equivalents can easily be derived for such elements as well. As an example, consider the NMOS transistor shown in Fig. 4.12(a). Its circuits representation is a non-linear VCCS controlled by both V_{GS} and V_{DS} (source to substrate voltage is neglected for simplicity). The values $V_{GS}^{(j)}$ and $V_{DS}^{(j)}$ are found for the *j*th iteration doing a Taylor series expansion about $V_{GS}^{(j-1)}$, $V_{DS}^{(j-1)}$.

$$f(V_{GS}^{(j)}, V_{DS}^{(j)}) = f(V_{GS}^{(j-1)}, V_{GS}^{(j-1)})$$

$$+ (V_{GS}^{(j)} - V_{GS}^{(j-1)})\frac{\partial f}{\partial V_{GS}} + (V_{DS}^{(j)} - V_{DS}^{(j-1)})\frac{\partial f}{\partial V_{DS}}$$

Fig. 4.12   (a) MOSFET (b) Its circuit model (c) Linearised equivalent, Note that the MOSFET is represented as a non-linear VCCS with more than one controlling voltage. In (c) the VCCS $\frac{\partial f}{\partial V_{DS}} V_{DS}^{(j)}$ is really a resistor of value $1 / \frac{\partial f}{\partial V_{DS}}$. A resistor can be thought of as a VCCS which has its own voltage as the controlling voltage.

where the partial derivatives are evaluated at $V_{GS}^{(j-1)}$, $V_{DS}^{(j-1)}$. The linearised equivalent corresponding to this is shown in Fig. 4.12(c). It consists of two linear VCCS in parallel with an independent current source. The VCCS $\partial f / \partial V_{DS}$ is actually a resistor. A resistor can always be thought of as a special kind of VCCS where the controlling voltage is the voltage across the element itself. It must be remembered that these more complicated linearised equivalents are difficult to find for user defined non-linear elements. They are more useful for built-in models like a model for a MOSFET incorporated in the package or program.

4.6 Summary

The DC solution of a non-linear circuit amounts to solving a system of simultaneous non-linear algebraic equations. The common technique used for this is the Newton-Raphson technique. A circuit interpretation can be given to this technique. This reduces the problem to that of solving a linear circuit at each Newton-Raphson iteration. The circuit interpretation simplifies the formulation of equations and automatically calculates only those derivatives that are needed at each iteration. The *pn* junction diode, though a very common circuit element, has to be carefully modelled to ensure proper convergence of the Newton-Raphson technique.

References

1. J.R. Rice, Numerical Methods, Software, and Analysis, Chap. 8, McGraw-Hill International Book Company, 1983.
2. L.W. Negel, SPICE 2: A Computer Program to Simulate Semiconductor Circuits, Ph.D. Thesis, University of California, Berkeley, May, 1978.
3. L.O. Chua and P.M. Lin, Computer-Aided Analysis of Electronic Circuits, pp. 224-227, Prentice Hall Inc., 1975.
4. N. Piskunov, Differential and Integral Calculus, pp. 273-275, Peace Publishers, Moscow.

Problems

1. Find the voltage V_d across the diode in the circuit shown. The diode is represented by the piecewise linear model, also shown. Do the analysis first by directly applying the Newton-Raphson method. Then repeat it by using the linearised equivalent for the diode. In both cases, assume an initial value for V_d equal to 1 V.

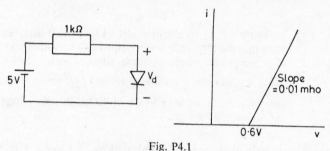

Fig. P4.1

2. The system of non-linear equations given below has four pairs of solutions. Find them all using the N-R technique with different guesses.

$$x_1 + 2x_2^2 = 1; \quad x_2 + 2x_1^2 = 1$$

3. Let the Newton-Raphson technique be used to solve a linear circuit. Show that convergence takes place in one step, no matter what the initial guess.

4. Solve for V_0 in each of the NMOS inverters shown. The load (upper) transistort in each case can be considered as a non-linear resistor. The drain to source current I_{DS} is given by

$$I_{DS} = 0 \quad \text{for} \quad V_{GS} \text{ less than } V_T$$

$$= \beta[(V_{GS} - V_T) V_{DS} - V_{DS}^2/2] \quad \text{for} \quad V_{DS} < (V_{GS} - V_T)$$

$$\text{and} \quad V_{GS} > V_T$$

$$= \frac{\beta}{2}(V_{GS} - V_T)^2 \quad \text{for} \quad V_{DS} \geqslant V_{GS} - V_T) \quad \text{and} \quad V_{GS} > V_T$$

For the driver transistors $V_T = 2$ V and $\beta = 5 \times 10^{-5}$ A/V².
For the load transistors, $V_T = -2$ V and $\beta = 10^{-5}$ A/V².

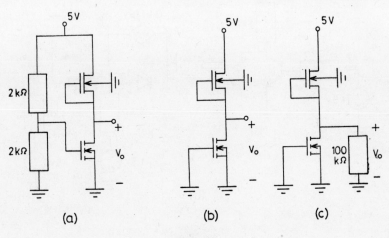

Fig. P4.4

5. Consider the CMOS inverter shown in Fig. P4.5. Equations governing the NMOS transistor are as in problem 4. A complementary set of relations apply for the PMOS transistor. Let the transconductance β of both transistors be 10^{-5} A/V². V_T for the NMOS transistor is 1.5 V and that for the PMOS transistor -1.5 V. Find V_0 by assuming an initial guess of (a) 1 V and (b) 4 V. Why are the answers different in the two cases? Use linearised equivalents assuming both transistors to be non-linear resistors. V_{GS} is constant for both transistors.

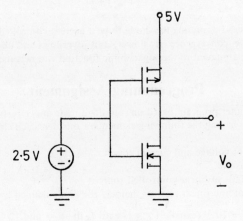

Fig. P4.5

6. Derive linearised equivalents corresponding to the N-R method for (a) non-linear CCCS (b) non-linear VCVS.

7. Solve for V_0 in each of the circuits given below assuming the diode to have a voltage controlled i-v relation

$$Id = 10^{-14} (\exp V_d/V_T - 1)A$$

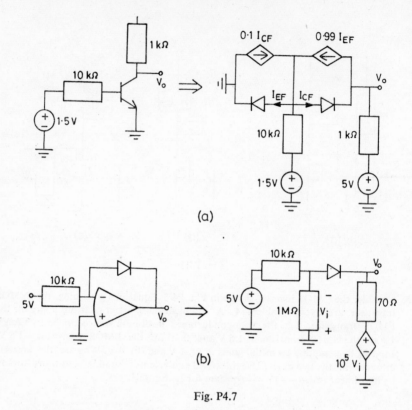

Fig. P4.7

8. Repeat problem 1 assuming the diode to be a current controlled non-linear element Note that the convergence is slower, but overflows and underflows are absent. However, the argument of the logarithmic function can become negative.

Programming Assignment

1. Write a computer program based on modified node analysis to analyse DC non-linear circuits having the following elements. Consider each element as a separate branch.
   1. Independent voltage and current sources
   2. Linear resistors
   3. All four kinds of linear controlled sources
   4. One kind of non-linear resistor, namely the diode defined by

   $$I_d = 10^{-14} (\exp V_d/V_T - 1) \text{ with } I_d \text{ in amps and } V_d \text{ in volts}$$

Use the idea of linearised equivalent for the diode. Modify the diode model suitably to avoid overflows and underflows. You may test your program initially for a piecewise linear model for the diode. Your program should clearly identify the model for the diode. You may use a program like the one at the end of Chapter 2 as a subroutine for solving linear circuits at each iteration.

Test your program for the two circuits given in Problem 7.

5

Transient Analysis of Linear and Non-linear Circuits

Transient analysis involves solving a network having independent sources which are arbitrary functions of time. So far we have looked only at DC independent sources. We now extend our treatment to include sources not necessarily DC. Transient analysis is inherently far more complex as we have to deal with differential equations as opposed to algebraic equations. Once again, it is important to distinguish between linear and non-linear circuits. Naturally, one has more options when solving linear circuits. For example, the Laplace transform technique is applicable only to linear circuits and helps in reducing differential equations to algebraic equations. Integration is done at the end when finding the inverse Laplace transform. Another popular approach is the state variable approach, applicable to both linear and non-linear circuits. It leads to a set of first order ordinary differential equations. In this chapter we mainly deal with a method derived, in a sense, from the state variable approach.

We now have to deal with two additional circuit elements, namely, inductors and capacitors. As defined in Chapter 1, these can be linear or non-linear. Methods to formulate equations have to take their i–v characteristics into account. Of special interest is sinusoidal small signal steady state analysis. Here, the exciting functions are sinusoidal and the circuit linear. In the steady state, such circuits can be analysed, without resorting to differential equations, using the concept of phasors. This, of course, is well known and is briefly described in the following section.

5.1 Sinusoidal Small Signal Linear Steady State Analysis

Small signal analysis is often done for amplifiers, active filters, etc. These circuits, in fact, go by the name of linear circuits though they use highly non-linear elements like transistors. This is because they are biased at some DC operating point around which a low amplitude (small) signal is applied. In such situations, a linear model at the operating point is constructed and its response to the small signal evaluated. For example a transistor may be replaced by its hybrid–π model [1] and its frequency response as an amplifier calculated. The analysis done is sinusoidal steady state analysis. It is the same as that done for AC circuits which have sinusoidal sources. It is assumed that the sources have been on for a long time so that transients have died down. This condition applies after a period equal to a few time constants of the circuit. What one is really doing in sinusoidal steady state analysis is solving the circuit in the Fourier transform domain.

In finding the steady state response of a linear circuit to sinusoidal sources, it is important that all sources be at the same frequency. One uses the generalised concept of impedance. Capacitors are replaced by an imaginary impedance $1/j\omega C$ and, inductors by $j\omega L$. Here ω is the angular frequency. While we deal with real resistances in DC analysis, complex impedances are used here. Independent sources are also complex. They are denoted by phasors having both magnitude and phase. So are the various other voltages and currents in the circuit. Once this is done, formulations like node analysis and loop analysis can be used. One arrives at a set of complex, algebraic linear equations. Solving n complex equations is equivalent to solving $2n$ real equations when the real and imaginary parts are separated out. The analysis is done at a particular frequency. When the frequency response is needed it is repeated at the various frequencies.

Sinusoidal steady state analysis, strictly speaking, is not part of transient analysis. The term 'transient analysis' is usually used for non-sinusoidal excitations. Sinusoidal steady state analysis uses the same techniques as DC linear analysis and could have been dealt with in Chapter 2 as well.

5.2 Possible Formulation Techniques

From the discussions in the past few chapters it is clear that the solution of a network proceeds in two stages.

1. Formulation of relevant algebraic or differential equations.
2. Mathematical solution of these equations.

Transient analysis leads to a set of non-linear differential equations, in general. Once these equations are arrived at, one could conceivably use some standard numerical routine to solve these equations. However, formulation of these equations is in itself not so straightforward.

Consider the circuit shown in Fig. 5.1. Let us try to use node analysis. We get

$$\frac{v_A - e(t)}{R_1} + \frac{1}{L}\int_{-\infty}^{t} v_A(\tau)\,d\tau + \frac{v_A - v_B}{R_2} = 0$$

$$\frac{v_B - v_A}{R_2} + C\frac{dv_B}{dt} = 0$$

These are integro-differential equations. One could differentiate the first equation to get a pure differential equation. But this is not an operation, a computer can easily perform. If mesh analysis were to be used, then capacitor currents would give rise to integrals.

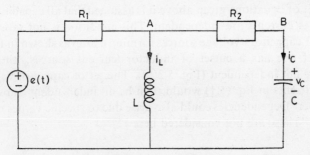

Fig. 5.1   Node analysis of this simple circuit leads to
integro-differential equations.

An obvious solution is to treat both capacitor voltages and inductor currents as unknowns. This is what is done in the state variable formulation. A set of first order differential equations are arrived at in terms of capacitor voltages and inductor currents unknowns. For example, let $(v_{C1}, v_{C2} \ldots v_{Ck})$ be the k capacitor voltages and $(i_{L1}, i_{L2} \ldots i_{Lm})$ the m inductor currents. The state variable formulation then gives

$$\frac{dv_{C1}}{dt} = f_{C1}(v_{C1}, v_{C2} \cdots v_{Ck}, i_{L1} \cdots i_{Lm})$$

$$\begin{array}{c} \cdot \\ \cdot \\ \cdot \end{array}$$

$$\frac{dv_{Ck}}{dt} = f_{Ck}(v_{C1}, v_{C2} \cdots v_{Ck}, i_{L1} \cdots i_{Lm})$$

$$\frac{di_{L1}}{dt} = f_{L1}(v_{C1}, v_{C2} \cdots v_{Ck}, i_{L1} \cdots i_{Lm})$$

$$\begin{array}{c} \cdot \\ \cdot \\ \cdot \end{array}$$

$$\frac{di_{Lm}}{dt} = f_{Lm}(v_{C1}, v_{C2} \cdots v_{Ck}, i_{L1} \cdots i_{Lm}) \qquad (5.1)$$

The various functions f_C and f_L are, in general, functions of the exciting sources and their derivatives. The state variable formulation applied to the circuit of Fig. 5.1 gives

$$\dot{v}_C = \frac{1}{C(R_1 + R_2)}(-v_C - R_1 i_L + e(t))$$

$$i_L = \frac{1}{L(R_1 + R_2)}(R_1 v_C - R_1 R_2 i_L + R_2 e(t))$$

The state variable formulation is equally applicable to linear and non-linear circuits. The functions f_C and f_L would accordingly be linear on non-linear. The set of equations given above can be solved using standard numerical techniques. A summary of these is given in the next section.

In the set of equations given above it is assumed that all capacitor voltages and inductor currents are independent. This condition is not satisfied if one has a set of capacitors/voltage sources forming a loop as shown in Fig. 5.2(a). Similarly, if one has a cutset of inductors/current sources, the inductor currents are not independent (Fig. 5.2(b)). The set of capacitor voltages and inductor currents in Eq. (5.1) would then be an independent subset of the total number. Dependencies could also arise due to specific values of controlled sources. These are not considered here [2].

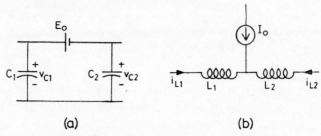

(a)   (b)

Fig. 5.2   (a) Circuit with loop of capacitors and independent voltage sources (b) Circuit with cutset of inductors and independent current sources.

The main problem with the state variable formulation is that it requires considerable matrix manipulation [2]. Rules to arrive at the set of equations in Eq. (5.1) are not easily stated and involve concepts not introduced in this book. Further one is forced to deviate from a circuit representation which we have tried to retain throughout our discussions so far. We will introduce an alternative formulation which requires no special matrix manipulations. It reduces transient analysis to the solution of DC networks. We have already developed techniques for both linear and non linear DC analysis. Above all, the formulation is equally general and a lot simpler.

This formulation is best explained using the Tableau method through it is not necessary that the Tableau method be used. In Sec. 2.2.2, it was pointed out that the last set of equations correspond to the branch constitutive relations for each branch or element. The first two sets are strictly a function of the topology. For a linear capacitor C the branch constitutive relation would be $C\,(dv_C/dt) - i_C = 0$. This differential equation is to be solved using some numerical technique. The Backward-Euler technique would give

$$v_{Cn+1} = v_{Cn} + \frac{dv_C}{dt_{/t=t_{n+1}}}\,\Delta t$$

It is assumed that values v_{Cn}, i_{Cn} at time step t_n are known. Values at time step $t_{n+1}\,(=t_n + \Delta t)$ are being evaluated. Rewriting the above expression, after substituting for dv_C/dt, we get

$$v_{Cn+1} = v_{Cn} + i_{Cn+1}\frac{\Delta t}{C}$$

The circuit representation of this equation is as shown in Fig. 5.3(a). In other words, at each time step, transient analysis amounts to solving a DC network with each capacitor replaced by the model of Fig. 5.3(a). Similarly the linear inductor equation, using the Backward-Euler integration scheme, is

$$i_{Ln+1} = i_{Ln} + v_{Ln+1}\frac{\Delta t}{L}$$

This equation corresponds to Fig. 5.3(b).

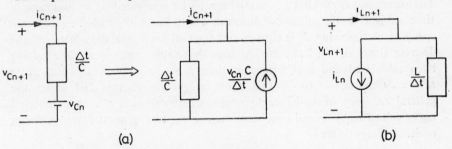

$$(a) \qquad\qquad\qquad\qquad\qquad\qquad (b)$$

Fig. 5.3 (a) Equivalent DC circuits at each time step for a linear capacitor using the Backward Euler integration scheme; (b) Equivalent DC circuit for a linear inductor using the Backward Euler integration scheme.

If the Tableau method is being used, then all inductor currents and capacitor voltages automatically become variables. An integration scheme has to be chosen and accordingly at each time step, inductors and capacitors are to be replaced by their associated model. The resulting DC network is to be solved at each time step. The associated model can be used even if the Tableau method of formulation is not directly used. Actually, node analysis, MNA, state variable analysis, etc., are all special cases of the Tableau method. Imagine, for example, that MNA is being used. One can still replace inductors and capacitors by their associated models and carry out the analysis. However, an additional rule has to be stated regarding the choice of current variables. All inductor currents must also be chosen as variables. These currents are needed to calculate parameters of the associated models. Capacitor voltages, being branch voltages, are not directly used in MNA. But it is very easy to calculate the branch voltage from the two node voltages. We have introduced the method of associated models using linear capacitors and inductors. It is equally applicable to non-linear capacitors and inductors. However, before considering these, a review of techniques to solve ordinary differential equations is necessary.

5.3 Numerical Solution of Ordinary Differential Equations

We will deal with systems of first order ordinary differential equations only. This is partly because the formulations used produce first order ODEs and partly because higher order equations can be reduced to first order equations [3]. Consider a set of equations of the form

$$\dot{x}_1 = f_1(x_1, x_2, x_3, \ldots, x_n, t)$$
$$\dot{x}_2 = f_2(x_1, x_2, x_3, \ldots, x_n, t)$$
$$\vdots$$
$$\dot{x}_n = f_n(x_1, x_2, x_3, \ldots, x_n, t)$$

The variables $(x_1(t), x_2(t) \ldots x_n(t))$ are to be evaluated as a function of time subject to the initial condition at $t = 0$ i.e. $x_1(0), x_2(0) \ldots x_n(0)$ are specified. A time step Δt is chosen and the various x_i are calculated at the discrete times $0, \Delta t, 2\Delta t$, etc. Assume that values upto some time t_n have been calculated and that values at time t_{n+1} are now to be found. Note $t_n = n\Delta t$ and $t_{n+1} = (n + 1)\Delta t$. Most integration schemes fall under the general category of multi-step routines where values at t_{n+1} are calculated using values from several time steps before t_n. The general formula for a multi-step algorithm is

$$x_i(t_{n+1}) = a_0 x_i(t_n) + a_1 x_i(t_{n-1}) + \ldots a_k x_i(t_{n-k})$$
$$+ \Delta t[b_{-1} f_i([x(t_{n+1})], t_{n+1}) + b_0 f_i([x(t_n)], t_n)$$
$$+ \ldots b_k f_i([x(t_{n-k})], t_{n-k})] \qquad (5.2)$$

The various constants a_i and b_i are found by matching the coefficients of a polynomial of some degree of order m[4]. Various integration algorithms emerge as special cases with suitable values for a_i and b_i. Some of these are

1. Forward Euler

$$x_i(t_{n+1}) = x_i(t_n) + \Delta t f_i([x(t_n)], t_n)$$

2. Backward-Euler

$$x_i(t_{n+1}) = x_i(t_n) + \Delta t f_i([x(t_{n+1})], t_{n+1})$$

3. Trapezoidal

$$x_i(t_{n+1}) = x_i(t_n) + \frac{\Delta t}{2}[f_i([x(t_{n+1})], t_{n+1}) + f_i([x(t_n)], t_n)]$$

4. Fourth Order Adams-Moulton

$$x_i(t_{n+1}) = x_i(t_n) + \frac{\Delta t}{24}[9f_i([x(t_{n+1})], t_{n+1}) + 19f_i([x(t_n)], t_n)$$
$$- 5f_i([x(t_{n-1})], t_{n-1}) + f_i([x(t_{n-2})], t_{n-2})]$$

While higher orders should, in principle, lead to better accuracy for a given time step Δt round off and truncation errors may make it pointless. The various multistep schemes can be classified as implicit or explicit depending on whether b_{-1} is non-zero or zero. Of the four algorithms listed above, the Forward-Euler scheme alone is explicit. In the others the quantity $x_i(t_{n+1})$ occurs on both sides of the equation i.e., it is not explicitly specified. Some iterative scheme must be used to solve for $x_i(t_{n+1})$ if f_i is non-linear. Explicit integration schemes are therefore simpler. However, they are inherently unstable as shown below:

Consider the differential equation in one unknown x

$$\frac{dx}{dt} = f(x, t)$$

Assume that we have solved for x upto some time t_n. The calculated value of x at $t = t_n$ is x_n, say. Then the calculated value x_{n+1} by Forward-Euler is

$$x_{n+1} = x_n + f(x_n, t_n) \Delta t$$

Let the calculated value of x_n have an error Δx_n. Then x_{n+1} has two sources of error

1. Δx_n due to error in x_n

2. $\dfrac{\partial f}{\partial x_{/x_n, t_n}} \Delta t \Delta x_n$ due to error in $f(x_n, t_n)$

Total error $\Delta x_{n+1} = \Delta x_n \left(1 + \dfrac{\partial f}{\partial x} \Delta t\right)$.

In the next step,

$$\Delta x_{n+2} = \left(1 + \frac{\partial f}{\partial x_{/t_{n+1}, x_{n+1}}} \Delta t\right) \Delta x_{n+1}$$

$$= \left(1 + \frac{\partial f}{\partial x_{t_{n+1}, x_{n+1}}} \Delta t\right)\left(1 + \frac{\partial f}{\partial x_{/t_n, x_n}} \Delta t\right) \Delta x_n$$

If $\left| + \dfrac{\partial f \Delta t}{\partial x} \right|$ is > 1.0, the error can get very large after a few time steps.
Consider the simple case

$$\frac{dx}{dt} = -\lambda x \text{ (positive)}$$

Now $\dfrac{\partial f}{\partial x} = \text{constant} = -\lambda$. The error, after m time steps from t_n, will be

$$\Delta x_{n+m} = (1 - \lambda \Delta t)^m \Delta x_n$$

If $\lambda \Delta t > 2$, error would be large. That is, if the time step Δt is greater than $2/\lambda$ the solution can become unstable. The solution is $x(t) = e^{-\lambda t}$, and for large t the solution should tend to zero. Instead it may tend towards infinity.

This instability has been illustrated with the simplest of the explicit algorithms. But all of them are potentially unstable. On the other hand consider the Backward-Euler and Trapezoidal implicit algorithms. Now

$$\Delta x_{n+1} = \frac{\Delta x_n}{\left(1 - \dfrac{\partial f}{\partial x_{/x_{n+1}, t_{n+1}}} \Delta t\right)} \qquad \text{(Backward Euler)}$$

$$\Delta x_{n+1} = \frac{\Delta x_n \left(1 + \frac{1}{2} \Delta t \, \frac{\partial f}{\partial x / x_n, t_n}\right)}{\left(1 - \frac{1}{2} \Delta t \, \frac{\partial f}{\partial x / x_{n+1}, t_{n+1}}\right)} \quad \text{(Trapezoidal)}$$

As long as $\partial f / \partial z$ is negative both the expressions above have a denominator greater than 1.0. Even for large Δt, Backward Euler and Trapezoidal algorithms will not blow up when solving $dx/dt = -\lambda x$. One may ask what happens if $\partial f / \partial x$ is positive and large. In that case the actual solution itself tends towards infinity and any integration algorithm would fail at some point.

In any multistep algorithm, values of x_i for a few time steps preceding the present one are needed. How then does one start? Usually some algorithm which does not require any preceding points is used. For example, if a fourth order Adams-Moulton algorithm were being used, $x_i(\Delta t)$ and $x_i(2\Delta t)$ must first be calculated using Forward-Euler, Runge-Kutta [4] or some other algorithm. From $t = 3\Delta t$ onwards fourth order Adams-Moulton can be used. In the discussion so far, the step size Δt has been taken as constant. Often, the algorithm can be designed to automatically adjust for the correct step size. In Predictor-Corrector algorithms, the new value is predicted using a relatively simple technique like Forward Euler. It is then compared with the actual value from the algorithm being used. If the two values are too far apart, the step size is halved. If they are too close, it is doubled. Otherwise it is left unchanged.

Sometimes the circuit being analysed may have two or more vastly different time constants. Such differential equations are called stiff differential equations. Co-efficients of the general multistep algorithm may be chosen according to Gear's algorithm [5] to solve such equations. Details are available in Ref. [4].

5.4 Associated Circuit Models for Inductors and Capacitors

Figure 5.3 shows the associated circuit models for a linear capacitor and a linear inductor corresponding to the Backward Euler algorithm. The equations corresponding to the Trapezoidal algorithm are

$$v_{Cn+1} = v_{Cn} + \frac{\Delta t}{2C} i_{Cn} + \frac{\Delta t}{2C} i_{Cn+1}$$

$$i_{Ln+1} = i_{Ln} + \frac{\Delta t}{2L} v_{Ln} + \frac{\Delta t}{2L} v_{Ln+1}$$

The corresponding models are given in Fig. 5.4. The concept of associated circuit models is easily extended to non-linear capacitors and inductors. These, naturally, lead to non-linear resistors in their circuit models.

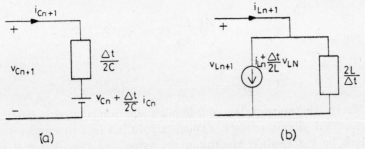

Fig. 5.4 Equivalent DC circuits at each time step corresponding to the Trapezoidal integration scheme for (a) linear capacitor; (b) linear inductor.

The effective inductance $L_{\text{eff}}(= d\lambda/dI)$ of a non-linear inductor can be represented as follows for the Backward-Euler algorithm

$$v_L = L_{\text{eff}}(i_L) \frac{di_L}{dt}$$

$$i_{Ln+1} = i_{Ln} + \frac{di_L}{dt_{/t_{n+1},\ i_{Ln+1}}} \Delta t$$

$$= i_{Ln} + \frac{\Delta t\ v_{Ln+1}}{L_{\text{eff}}\ (i_{Ln+1})}$$

This represents a non-linear resistor whose i–v relationship is given by

$$I = A + \frac{BV}{L_{\text{eff}}(I)}$$

where A and B are constants. As a second example, consider the transition capacitance of a pn junction diode given by

$$C_T(v) = \frac{C_0}{(1 - v)^{1/2}}$$

By the trapezoidal algorithm

$$v_{Cn+1} = v_{Cn} + \frac{\Delta t}{2}\left[\frac{dv_c}{dt_{/v_{Cn},t_n}} + \frac{dv_c}{dt_{/v_{Cn+1},t_{n+1}}}\right]$$

$$= v_{Cn} + \frac{\Delta t}{2}\left[\frac{i_{Cn}}{C_T(v_{Cn})} + \frac{i_{Cn+1}}{C_T(v_{Cn+1})}\right]$$

or,

$$v_{Cn+1} = v_{Cn} + \frac{\Delta t}{2}\frac{i_{Cn}}{C_T(v_{Cn})} + \frac{\Delta t}{2}\frac{i_{Cn+1}}{C_T(v_{Cn+1})}$$

This corresponds to a non-linear resistor whose i–v characteristics is given by

$$V = A + BI(1 - V)^{1/2}$$

where

$$A = v_{Cn} + \frac{\Delta t}{2} \frac{i_{Cn}}{C_0} (1 - v_{Cn})^{1/2}$$

and

$$B = \frac{\Delta t}{2C_0}$$

The difference between explicit and implicit algorithms is also reflected in the associated circuit models. Explicit algorithms lead to simpler models. Whatever be the explicit algorithm, the associated circuit model for an inductor (non-linear or linear) is an independent current source. Similarly any explicit algorithm gives an independent voltage source for a capacitor (non-linear or linear) associated circuit model. For implicit algorithms, a general inductor gives a non-linear current controlled resistor as the associated circuit model and a general capacitor, a non-linear voltage controlled resistor.

Some considerations in choosing a suitable integration algorithm and equation formulation scheme can now be stated. Node analysis cannot be used unless there are no inductors in the circuit. MNA is alright, provided inductor currents are chosen as variables. Of course, the Tableau method is most general and can certainly be used. The formulation can to some extent limit the choice of integration algorithms. Backward Euler is the only implicit algorithm which does not require one to calculate capacitor currents explicitly. It is simple and relatively stable. As node analysis and MNA may not provide capacitor currents without extra work, Backward Euler is very convenient for use with node analysis and MNA. Higher order integration algorithms will need currents and voltages from many previous time steps. In such cases, starting values have to be obtained from lower order schemes.

5.5 Use of Associated Circuit Models

The step for doing transient analysis using associated circuit models can be summarised as:

1. Choose some integration algorithm, preferably an implicit one. Also choose a time step Δt.
2. Replace every inductor and capacitor at each time step by an associated circuit model corresponding to the integration scheme. Non-linear inductors and capacitors lead to non-linear resistors.
3. Solve the resulting DC network (at each time step) using some formulation like MNA. If the resulting network is non-linear, the solution would have to be done iteratively as described in Chapter 4.
4. Increment time and repeat steps 2 and 3 till the solution is available for the entire time interval specified.

The initial conditions that need to be specified at $t = 0$ are the capacitor voltages and inductor currents. Transient analysis first requires a DC analysis at $t = 0$. Non-linear transient analysis requires one to solve a linear DC circuit many, many times. First a solution has to be obtained at every time step. Secondly, at every time step several N–R iterations have to be done to solve the DC non-linear network resulting from the substitution of associated circuit models. Fortunately, the number of N–R iterations at each time step will not be large. This is because a good initial guess can be made. This guess could just be values from the previous time step. Some trade-off can be made between the number of N–R iterations at each time step and the number of time steps.

Example 5.1: Consider the simple RC circuit shown in Fig. 5.5(a). The input $e(t)$ is a step function. It is required to find $v_C(t)$, for $t > 0$ with the condition $v_C(0) = 0$ V.

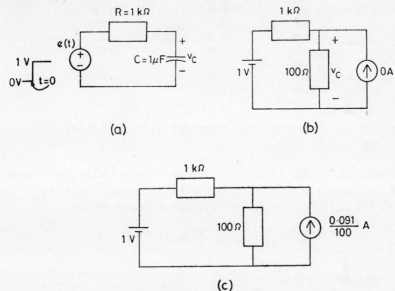

Fig. 5.5   (a) Transient analysis for a simple RC network; (b) DC equivalent at $t = \Delta t$ (c) DC equivalent at $t = 2\Delta t$.

First, we choose an integration scheme. The Backward Euler scheme is a convenient and effective choice. The RC time constant (1 K ohm × 1 microfarad) is 1 msec. We choose Δt equal to 100 μ sec. This is on the large side, but our interest, here, is to only illustrate the use of associated circuit models. From Fig. 5.3(a), the circuit model for a capacitor is a resistor $\Delta t/C$ in parallel with a current source $v_{Cn}/(\Delta t/C)$.

$$\frac{\Delta t}{C} = \frac{100 \ \mu \text{ sec}}{1 \ \mu \text{F}} = 100 \text{ ohms}$$

At $t = \Delta t$, $v_{Cn} = v_C(0) = 0$ V. At $t = 2\Delta t$, $v_{Cn} = v_C(\Delta t)$ and so on. Corresponding associated circuit models are shown for $t = \Delta t$ and

$t = 2\,\Delta t$ in Fig. 5.5(b) and (c) respectively. Solving the resulting DC network at each value of t we get the values shown in Table 5.1. The exact values are also shown for comparison. The exact solution is

$$v_C(t) = 1\,(-\,e^{-t/RC}).$$

Table 5.1

t (μ secs)	v_C from Backward Euler (in V)	v_C from exact solution (in V)
100	0.091	0.095
200	0.174	0.1813
300	0.249	0.259
400	0.317	0.330
500	0.379	0.394
600	0.435	0.451
700	0.477	0.503
800	0.524	0.551
900	0.567	0.593
1000	0.606	0.632

The time step chosen was large. Otherwise the results would have been more accurate.

Example 5.2: In the diode circuit shown in Fig. 5.6(a), $e(t)$ is switched suddenly to $-5\,V$ after it has been at 0 V for a long time. For $t > 0$, the diode can be replaced by a non-linear transition capacitance

$$C_{\text{eff}}\,(V) = \frac{10^{-9}}{(1 - V)^{1/2}}\cdot \text{ Find } v(t) \text{ for } t > 0.$$

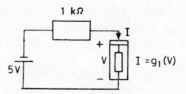

Fig. 5.6 (a) Charging in reverse bias of transition capacitance of a *pn* junction diode; (b) circuit equivalent (c) DC equivalent corresponding to Backward Euler.

The capacitance value is $\geqslant 10^{-9}\, F$. So $RC \geqslant 1\, \mu$ sec. Let us choose a time step $\Delta t = 50$ nsec and the Backward Euler integration algorithm. $v(0) = 0$ V. As was shown in the last section, the non-linear capacitance is to be replaced by a non-linear resistance.

$$v_{Cn+1} = v_{Cn} + \frac{\Delta t}{C(v_{Cn+1})}\, i_{Cn+1}$$

$$= v_{Cn} + \frac{\Delta t}{10^{-9}}\, i_{Cn+1}\, (1 - v_{Cn+1})^{1/2}$$

or,

$$i_{Cn+1} = \frac{v_{Cn+1} \times 10^{-9}}{(1 - v_{Cn+1})^{1/2} \Delta t} - \frac{v_{Cn} \times 10^{-9}}{\Delta t (1 - v_{Cn+1})^{1/2}} \tag{5.3}$$

At $t = \Delta t$ the capacitor is replaced by a non-linear resistor given by substituting $v_{Cn}\ (= v(0))$ in the above expression. This gives

$$I = g_1(V) = \frac{10^{-9}\, V}{(1 - V)^{1/2} \times 50 \times 10^{-9}}$$

as the i–v relation. The circuit of Fig. 5.6(c) must now be solved with the above i–v relation for the voltage controlled non-linear resistor. Let the initial guess $V_1^{(0)} = 0$ V. The subscript 1 refers to time step 1 or the time $t = \Delta t$

$$g_1(0) = 0$$

$$\frac{dg_1}{dV} = \frac{1}{50(1 - V)^{1/2}} + \frac{V}{50(1 - V)^{3/2}} \times \frac{1}{2}$$

$$\frac{dg_1}{dV}_{/V=0} = \frac{1}{50}\ \text{mhos}$$

Solving the circuit of Fig. 5.7(a) gives the first iteration $V_1^{(1)} = -0.238$ V.

$$g_1(-0.238\ \text{V}) = -4.278\ \text{mA}$$

$$\frac{dg_1}{d_1 V}_{/-0.238\ \text{V}} = 0.0180 - 0.0017 = 0.0163\ \text{mhos}$$

$$g_1(-0.238\ \text{V}) + 0.238\, \frac{dg_1}{dV}_{/-0.238\ \text{V}} = -4.278 \times 10^{-3} + 3.879 \times 10^{-3}$$

$$= -3.986 \times 10^{-4}\ \text{A}$$

The circuit to be solved is that shown in Fig. 5.7(b). Solving, we get

$$V_1^{(2)} = -0.266\ \text{V}$$

We stop at two N–R iterations and say v_C at $t = \Delta t$ is -0.266 V. At $t = 2\, \Delta t$ the non-linear resistor is given by substituting $v_{Cn} = -0.266$ V above in Eq. (5.3).

$$I = g_2(V) = \frac{10^{-9}\, V}{(1 - V)^{1/2} \times 50 \times 10^{-9}} + \frac{0.266 \times 10^{-9}}{50 \times 10^{-9} \times (1 - V)^{1/2}}$$

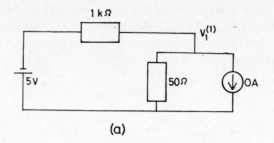

(a)

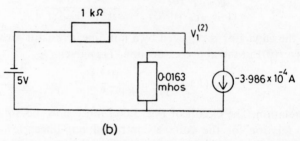

(b)

Fig. 5.7 At $t = \Delta t$ the non-linear DC equivalent of the non-linear capacitor is analysed using the Newton-Raphson technique (a) Linearised equivalent for 1st N-R iteration; (b) Linearised equivalent for 2nd N-R iteration.

Let the initial guess be

$$V_2^{(0)} = -0.266 \text{ V. Then } g_2(-0.266 \text{ V}) = -4.728 \text{ mA} + 4.728 \text{ mA} = 0$$

$$\frac{dg_2}{dV} = 1.778 \times 10^{-2}$$

Current source in parallel $= g_2 - V\dfrac{dg_2}{dV} = 0 + 0.266 \times 1.778 \times 10^{-2}$

$$= 4.730 \times 10^{-3} \text{ A}$$

The circuit to be solved is shown in Fig. 5.8(a). This gives

$$V_2^{(1)} = -0.518 \text{ V}$$

$$g_2(-0.518) = -4.09 \times 10^{-3} \text{ A}$$

$$\frac{dg_2}{dV_{/V=-0.518}} = 1.623 \times 10^{-2} - 1.347 \times 10^{-3} = 1.489 \times 10^{-2}$$

The circuit to be solved is shown in Fig. 5.8(b). Solving, we get

$$V_2^{(2)} = -0.543 \text{ V}$$

Stopping at 2 N-R iterations we get v_C at $t = 2 \Delta t$ is -0.543 V.

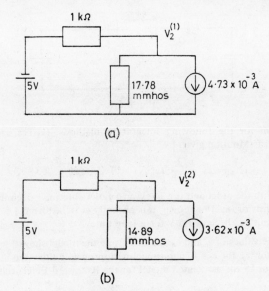

Fig. 5.8   Linearised equivalents at $t = 2\Delta t$ for (a) first
N-R iteration; (b) second N-R interation.

5.6 Summary

Non-linear transient analysis is computationally most expensive and difficult. The numerical operation involved is that of solving non-linear differential equations. Here again a circuit interpretation makes the formulation
and solution far easier. At each time step, one can find an associated model
for the capacitors and inductors in the circuit. In general, these models
contain non-linear resistors. Thus the problem is reduced to one of solving
a DC non-linear network at each time step. The associated model depends
on the integration scheme used. A convenient scheme for networks with
capacitors, but no inductors, is the Backward Euler integration scheme
which ties in well with node analysis for such networks.

References

1. J.F. Gibbons, Semiconductor Electronics, p. 499, McGraw-Hill, Inc., 1968.
2. L.O. Chua and P.M. Lin, Computer-Aided Analysis of Electronic Circuits,
   Chap. 8, Prentice-Hall, Inc., 1975.
3. J.H. Rice, Numerical Methods, Software and Analysis, pp. 266-267, McGraw-Hill
   International Book Company, 1983.
4. L.O. Chua and P.M. Lin, Computer-Aided Analysis of Electronic Circuits,
   Chap. 11, Prentice-Hall Inc., 1975.
5. C.W. Gear, Numerical Initial Value Problems in Ordinary Differential Equations,
   Prentice-Hall, Inc., 1971.

Problems

1. Solve for $v(t)$ in the simple RC circuit shown replacing C by the associated network
   and choosing an appropriate time step. Assume $v(0) = 0$ and solve for $t \geqslant 0$. Do

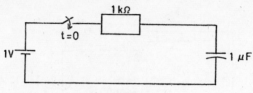

Fig. P 5.1

the problem for the following integration methods. (a) Trapezoidal (b) Fourth order Adamas-Moulton given by

$$x_{n+1} = x_n + \frac{\Delta t}{24}(9x'_{n+1} + 19x'_n - 5x'_{n-1} + x'_{n-2})$$

Compare with the exact and the Backward Euler solutions given in Example 5.1.

2. For the transistor inverter shown, v_i is abruptly switched from 1.5 V to 0 V at $t=0$ after the input has been at 1.5 V for a long time.

   (a) Find the values of v_0 and v_B at $t = 0$ using the model shown.
   (b) Solve for v_0 for $t \geq 0$ choosing an appropriate time step and replacing the capacitor by the associated model for the Backward-Euler technique.

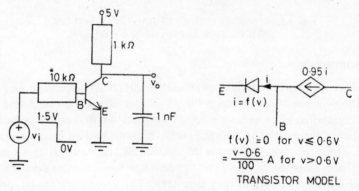

Fig. P 5.2

3. For the NMOS inverter shown v_{in} switches abruptly from 5 V to 0 V at $t = 0$. Here C_0 represents the net loading due to subsequent stages, stray capacitances, etc. Find $v_0(t)$ for $t \geq 0$.

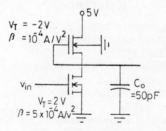

Fig. P 5.3

4. In the circuit shown the diode can be replaced by an ideal diode without any capacitance and two voltage dependent capacitances in parallel. These are $C_T = 10^{-9}/(1.0 - v_d)^{0.5}$ which is the transition or depletion capacitance and

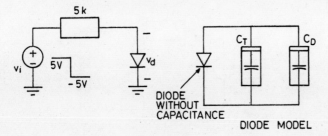

Fig. P 5.4

$C_D = C_{DO} f(v)$, where $f(v)$ is the diode equation. C_D is the diffusion capacitance. As an approximation, let the diode equation be replaced by the piecewise model shown in Problem 2. C_{DO} is so chosen that $C_D = 50$ nF when $v_d = 0.7$ V. When v_i is abruptly switched from 5 V to -5 V find v_d and i as a function of time. Note that C_T and C_D defined above are the effective capacitances $= dQ/dV$.

5. In solving an equation of the type $dx/dt = f(x, t)$ let the error at time t_n be Δx_n, Show that the error at time $t_{n+1} = t_n + \Delta t$ is

(a) $$\dfrac{\Delta x_n}{1 - \dfrac{\partial f}{\partial x}\Big|_{x_{n+1},\ t_{n+1}} \Delta t} \qquad \text{for the Backward-Euler method}$$

(b) $$\dfrac{\Delta x_n \left(1 + \dfrac{1}{2}\dfrac{\partial f}{\partial x}\Big|_{x_n,\ t_n} \Delta t\right)}{\left(1 - \dfrac{1}{2}\dfrac{\partial f}{\partial x}\Big|_{x_{n+1},\ t_{n+1}} \Delta t\right)} \qquad \text{for the Trapezoidal method}$$

6. The thyristor in Fig. P5.6. is fired at $\theta = 90°$. Considering it to be an ideal switch, calculate $i(t)$ replacing the inductor by the model corresponding to Trapezoidal method Calculate for one cycle.

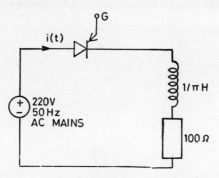

Fig. P 5.6

7. The non-linear capacitance used in the circuit of Fig. P5.7(a) has the variation shown in Fig. P5.7(b). A negative step is applied at $t = 0$ after the circuit has been in steady state. The value of v_0 is to be found at $t = \Delta t$, where Δt is 10 microseconds.

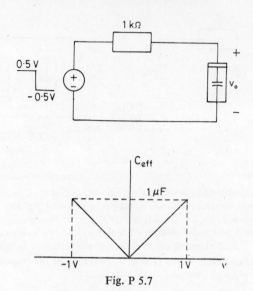

Fig. P 5.7

(a) Represent the capacitor by a non-linear resistor to solve for $v_0(t)$ at $t = \Delta t$. Use the Backward-Euler integration scheme.

(b) Solve the non-linear *DC* network of (a) using linearised equivalents. What would be the initial guess?

Programming Assignment

(1) Assume that the program in programming assignment 1 of Chapter 4 is available. Use that as a routine to do transient analysis. Only one kind of storage element, a linear capacitor is allowed. The independent *DC* sources can be current or voltage sources. The time-varying independent sources have to be piecewise linear voltage sources. Transient analysis should be preceded by *DC* analysis at $t = 0^-$. Verify your program with the circuit in Problem 2 above.

6

Models for Common Semiconductor Devices

Any circuit analysis program intended for design of electronic circuits must have built-in models for the common semiconductor devices used. The user himself may not be knowledgeable enough to think of a model for a device like, say, a MOSFET. Even if he is, user defined non-linear elements are difficult to handle unless they are defined by simple functions like polynomials or are piecewise linear. A built-in model can have the various necessary derivatives also built in. Linearised equivalents for non-linear DC analysis and associated circuit models for the inductors and capacitors are then easily constructed. Programs like SPICE [1] have built-in models for BJTs, MOSFETs, etc. A particular device can be simulated by feeding in appropriate parameters. For example all BJTs may be described by the Ebers-Moll Model. A particular BJT may be simulated by feeding in values of α, reverse saturation current, etc. It is a limitation for the user to stick to a built-in model. If the program uses the Ebers–Moll model and the user would like to use the transport model, there is a problem. Of course, the user can write his own program and put in whatever model he wants. Alternatively, he can build his own subroutine into an existing program.

The models described in this chapter are based on the models used in SPICE [1]. As this program has come to be accepted as a standard, it seems to be justified. In any case, extensions to these models and other references are given so that he knows how to describe a device in terms of the SPICE parameters. No circuit simulation is possible unless proper values are given for device parameters. The reader is expected to have some familiarity with the theory of semiconductor devices. The emphasis, in this chapter, is on models and no attempt is made to describe semiconductor theory.

6.1 Kinds of Models

A model is a rather vague term and one can talk of many kinds of models. The kind of models acceptable to us must have a circuit representation. The model may have elements like non-linear resistors or controlled sources. But it cannot be a model which is defined by a differential equation. The circuit elements of the model would be defined by, among other things, the kind of analysis required. DC analysis requires no inductors or capacitors.

Small signal analysis requires linear elements. A global model would be valid over all regions of operation. A local model for a BJT, say, may be valid only in the normal active region.

An attempt is made to include an AC global model in general circuit analysis programs and is designed to be valid for all possible values of voltages and currents. Further non-linear capacitors and inductors are used to describe transient behaviour. If one is not interested in certain features, corresponding parameters may be chosen to exclude these features. Programs have default values for these parameters and these values are so chosen that only the gross features are represented. For DC analysis, inductors and capacitors are automatically left out from the model. For small signal analysis, a linear incremental model is constructed at the operating point.

The models described in this chapter are physical models in the sense that they are based on physical processes going on inside the device. Sometimes the device may be taken as black box and the model arrived at from external measurements. The Thevenin equivalent of a network is one such example. Such black box models are not discussed in this book.

6.2 Model for *pn* Junction Diodes

A circuit model for a *pn* junction diode is shown in Fig. 6.1. The linear resistor R_s is the ohmic effect of the bulk of the neutral semiconductor outside the space charge or transition region. Also included in R_s are the contact resistances. The non-linear resistance represents the basic non-linear *i-v* characteristics of the diode defined by

$$I_D = I_S \left(\exp \left(\frac{V_D}{\eta V_T} \right) - 1 \right) \qquad (6.1)$$

Here,

I_S = Reverse saturation current.

$V_T = kT/q$, where k is the Boltzman constant (1.38×10^{-23} joules/°K).

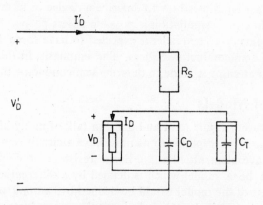

Fig. 6.1 Circuit model for a *pn* junction diode

T the temperature in °K and q the electronic charge in coulombs. At room temperature of 300 °K, $V_T = 26$ mV.

η = emission constant.

Figure 6.2 shows ln (I_D/I_S) as a function of V_D/V_T. For low currents the slope is about 0.5, then it increases to 1 and then again decreases to 0.5. At low currents, the dominant component is due to space charge recombination [2] while at higher currents it is due to diffusion [3]. Hence the change in slope. At very high currents, the slope again drops due to high injection effects. In germanium diodes, as opposed to silicon diodes, space charge recombination currents are small. Representing the entire region by one value of η is inaccurate for silicon transistors.

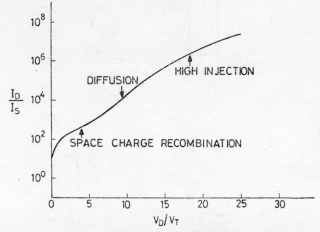

Fig. 6.2 Typical i–v relation for a pn junction Si diode. At low values of current $\eta \simeq 2$ due to current being dominated by space charge recombination. At moderate values of current $\eta \simeq 1$ as diffusion dominates. At high values of current η is again $\simeq 2$ because of high injection effects.

Figure 6.1 shows two non-linear capacitors C_D and C_T. C_D is the diffusion capacitance and C_T the transition or depletion region (also space charge region) capacitance. C_T is due to the exposed immobile charges (as shown in Fig. 6.3(b)) in the depletion or space charge region. Various expressions for the depletion region are given below and can be found in any standard text book (see, for example, [3], [4]).

$$N_A x_P = N_D x_n$$

$$X_d = x_n - x_p = x_n + |x_p| = \left(\frac{2\epsilon_{Si}}{q}\right)^{1/2} (\phi_0 - V)^{1/2} \left(\frac{N_A + N_D}{N_A N_D}\right)^{1/2}$$

$$E_0 = -\left(\frac{2q}{\epsilon_{Si}} \frac{N_A N_D}{N_A + N_D}\right)^{1/2} (\phi_0 - V)^{1/2}$$

$$Q_j = A \left[2\epsilon_{Si} q \frac{N_A N_D}{N_A + N_D}\right]^{1/2} (\phi_0 - V)^{1/2}$$

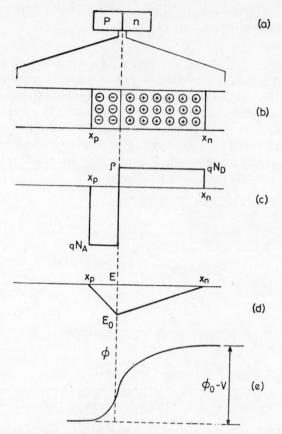

Fig. 6.3 Space charge region of a *pn* junction diode. (a) Diode (b) Immobile charges in space charge region (c) Charge distribution for step junction (d) Electric field variation. V is the externally applied bias and ϕ_0 the junction potential for zero bias

$$C_T = \frac{dQ_j}{dV} = A\left[2\epsilon_{Si}\, q\, \frac{N_A N_D}{N_A + N_D}\right]^{1/2} \frac{1}{2(\phi_0 - V)^{1/2}}$$

$$\phi_0 = V_T \ln\left(\frac{N_A N_D}{n_i^2}\right) \qquad\qquad (6.2)$$

In the above equations,

N_A, N_D = Acceptor and donor impurity levels in p and n regions respectively.

x_p, x_n = Extent of depletion region in p and n regions (Fig. 6.3(b) and (c)).

ϵ_{Si} = Permittivity of silicon (silicon device assumed).

E_0 = Maximum electric field (Fig. 6.3 (d)).

Q_j = Immobile charge on either side of junction.

ϕ_0 = Zero bias junction potential.

n_i = Intrinsic concentration.

One can rewrite the expression for C_T above as

$$C_T = \frac{C_{JO}}{(1 - V/\phi_0)^{1/2}}$$

where C_{JO} = zero bias depletion capacitance.

An abrupt function has been assumed in the expression given above. More generally

$$C_T = \frac{C_{JO}}{(1 - V/\phi_0)^m} \tag{6.3}$$

where m = grading coefficient.

For an abruptly graded junction $m = \frac{1}{2}$ and for a linearly graded junction $m = \frac{1}{3}$. It may be noted that Eq. 6.3 is accurate only for a reverse biased junction. Often the impurity concentration on one side of the junction far exceeds the other. For example, if $N_A \gg N_D$ the term

$$\frac{N_A N_D}{N_A + N_D} \simeq N_D$$

Then the depletion region exists mainly on the n side and the capacitance depends mainly on N_D.

The diffusion capacitance C_D is due to the build up of mobile charge immediately outside the depletion region as shown in Fig. 6.4(a). The figure pertains to conditions during forward bias in the steady state. For reverse bias the mobile charge is depleted as shown in Fig. 6.4(b). If a diode is abruptly switched from forward to reverse bias the excess mobile charge has to be removed. This takes a finite time and contributes to the diffusion capacitance. For a transition from reverse to forward bias, similarly, the mobile charge must build up, once again contributing to a capacitance.

Consider, first, a diode long enough so that the carrier levels decrease to the zero bias levels away from the junction but well before the contacts. Such a diode is a long base diode. The carrier concentration profiles and total excess charges are then given by [4].

$$n_p(x) - n_{po} = n_{po}(e^{V/V_T} - 1) \exp\left(- \mid (x - x_p) \mid /L_n\right)$$

$$\simeq n_{po}(e^{V/V_T} - 1) \exp\left(- \mid x \mid /L_n\right)$$

$$p_n(x) - p_{no} = p_{no}(e^{V/V_T} - 1) \exp\left(-(x - x_n)\right)/L_p$$

$$\simeq p_{no}(e^{V/V_T} - 1) \exp\left(-x/L_p\right)$$

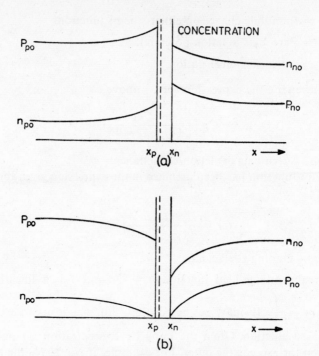

Fig. 6.4 Minority and majority carrier profiles near the junction
for (a) forward bias (b) reverse bias. Long based diode
is assumed.

$$Q_p = qA\int_{(x_n \simeq 0)}^{\infty} (p_n - p_{no})\, dx = qAL_p p_{no}(e^{V/V_T} - 1)$$

$$Q_n = -qAL_n n_{po}(e^{V/V_T} - 1)$$

$$Q_{\text{diff}} = Q_p + |Q_n| = qA(e^{V/V_T} - 1)[L_p p_{no} + L_n n_{po}]$$

$$C_D = \frac{dQ_{\text{diff}}}{dV} = \frac{qAe^{V/V_T}}{V_T}[L_p p_{no} + L_n n_{po}] \tag{6.4}$$

In the above expressions,

n_{po} = minority (electron) concentration far from junction in the p region.

p_{no} = minority (hole) concentration far from junction in the n region.

n_p, p_n = minority carrier concentrations in p and n regions respectively.

Q_{diff} = total charge contributing to the diffusion capacitances.

L_p, L_n = diffusion lengths for holes and electrons respectively.

A = area of diode.

Again considering the case where $N_A \gg N_D$, $p_{no} \gg n_{po}$ we get

$$C_D = \frac{qA}{V_T}e^{V/V_T} L_p p_{no}$$

If $p_{no} \gg n_{po}$ the reverse saturation current is given by

$$I_s \simeq qA \frac{D_p}{L_p} p_{no}$$

Substituting $\left(\frac{D_p}{L_p}\right)^{-1} = \tau_p$ we get

$$C_D = \frac{I_s}{V_T} \tau_p e^{V/V_T}$$

where $\tau_p =$ minority carrier lifetime in n region.

If $\eta \neq 1$ then

$$C_D = \frac{I_s}{\eta V_T} \tau_p e^{V/\eta V_T} \tag{6.5}$$

Consider, next, a diode with a width W such that $W \ll L_p$, shown in Fig. 6.5. Such a diode, called a short base diode, has a minority carrier concentration profile which is approximately linear. Again assuming $N_A \gg N_D$

$$p_n(x) - p_{no} = p_{no}(e^{V/V_T} - 1)\left(1 - \frac{x}{W}\right)$$

$$Q_{\text{diff}} = qA \int_{x_p}^{\infty} (p_n(x) - p_{no})dx = \frac{1}{2}qAW p_{no}(e^{V/V_T} - 1)$$

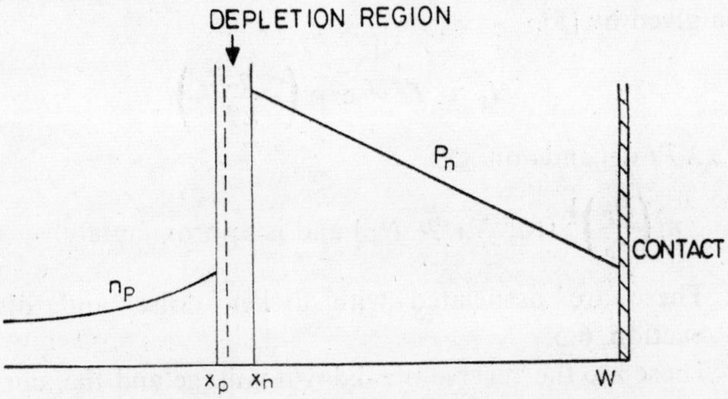

Fig. 6.5 Minority carrier profiles for a short based diode

The forward bias current is approximately equal to the slope of hole distribution in the region. So

$$I_D \simeq \frac{qAD_p p_{no}}{W}(e^{V/V_T} - 1)$$

which gives

$$I_s = \frac{qAD_p p_{no}}{W}$$

Defining a quantity $\tau_t = \frac{W^2}{2D_p}$ we get

$$Q_{\text{diff}} = \tau_t I_s \left(\exp\left(\frac{V}{V_T} - 1\right)\right)$$

and

$$C_D = \frac{I_s}{V_T} \tau_t \ \exp\left(\frac{V}{V_T}\right)$$

More generally,

$$C_D = \frac{I_s}{\eta V_T} \ \tau_t \exp\left(\frac{V}{\eta V_T}\right) \tag{6.6}$$

which is similar to Eq. (6.5) with τ_p replaced by τ_t, τ_t can be interpreted as the transit time across the width W.

In short, whether a diode is long based or short based, its diffusion capacitance is defined in terms of a lifetime or transit time τ. This is strictly true only when one side of the junction is far more highly doped. Otherwise, the expressions for C_D are more complicated. Short base diodes are present in transistors where the base is very thin and lightly doped compared to the emitter. In bipolar ICs, diodes are often realised by shorting the C–B junction of a transistor.

Some of the other parameters used in the SPICE model for a diode are:

EG : This is the activation energy and is the difference in energy between the valence and conduction bands. For Si, $EG = 1.11$ eV.

XTI : This is the saturation current temperature exponent. It can be shown that the reverse saturation current I_s has a temperature variation given by [5].

$$I_s \propto T^{XTI} \exp\left(\frac{-qEG}{kT}\right)$$

XTI depends on

$$n_i^2\left(\frac{D_p}{\tau_p}\right)^{\frac{1}{2}} \ \text{(for } N_A \gg N_D) \text{ and is approximately } = 3 \text{ for } S_i$$

KF, AF : These are associated with flicker noise and discussed in Section 6.5.

BV, IBV : These are the reverse breakdown voltage and the current at the reverse breakdown voltage respectively.

FC : The expression for C_T given in Eq. (6.3) is not quite valid for forward bias especially when V approaches ϕ_0. The factor *FC* modifies the expression for C_T during forward bias.

Table 6.1 gives the SPICE diode parameters [6] [4]. Parameters I_s, R_s and C_{JO} depend on the area of the device. While these parameters have been explained here keeping a *pn* junction diode in mind other diodes like Schottky diodes can also be represented by choosing parameters appropriately.

The model, though it has as many as 14 parameters, still has a number of limitations. Among these are

1. For reverse bias, the typical current is several orders of magnitude greater than I_s. This is due to space charge generation, leakage,

Table 6.1 Parameters for SPICE Diode Model

Name	Symbol Used	Description	Units	Default	Typical
IS	I_s	Reverse saturation current	A	1.0×10^{-14}	1.0×10^{-14}
RS	R_s	Bulk and contact resistance	Ohm	0	10
N	η	Emission coefficient	—	1.0	1.0
TT	$\tau_t, \tau_p, \tau_n,$	Transit time/Life-time	sec.	0	0.1 ns
CJO	C_{JO}	Zero bias depletion capacitance	F	0	2 pF
VJ	ϕ_0	Contact potential	V	1.0	0.8
EG	E_G	Activation energy	eV	1.11	1.11 (for Si)
M	m	Grading Coefficient	—	0.5	0.5
XTI	XTI	Saturating current temperature of exponent	—	3	3
KF	K_f	Flicker noise coefficient	—	0	
AF	A_f	Flicker noise exponent	—	1	
FC	FC	Coefficient for forward bias depletion capacitance	—	0.5	
BV	BV	Reverse breakdown voltage	V	∞	40.0
IBV	IBV	Current at break-down voltage	A	1 mA	

    defects, etc. Therefore, the reverse current given by the model is inaccurate.

2. As shown in Fig. 6.2, the diode current has different emission constants in different regions. It is not accurate to describe the entire region by one emission constant or coefficient.

3. Phenomena like tunnelling are not represented.

6.3 Models for Bipolar Junction Transistors (BJTs)

A simple, but very useful model for a BJT is the Ebers-Moll model [7]. It is given in virtually every introductory text book and is shown in Fig. 6.6. Two versions of the mode are shown and we will use that shown in Fig. 6.6(b) As the model shows, a BJT is just two diodes connected back with CCCS' to indicate currents transported across the thin base from collector to emitter and vice versa.

    Transistors are so designed that transport of carriers from emitter to collector is far more efficient than from collector to emitter. Therefore, while α_F is very close to 1 ($\simeq 0.99$) α_R is well under 0.5. Figure 6.6 shows the model for an *NPN* transistor. All polarities would reverse for a PNP

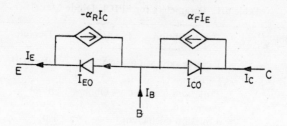

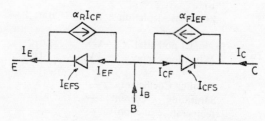

Fig. 6.6   (a) Ebers–Moll model for a BJT·I_{EO} and I_{CO} are the reverse saturation currents for the diodes shown. (b) Another representation of the Ebers-Moll model. I_{EFS} and I_{CFS} are the reverse saturation currents of the diodes shown.

transistor. If I_{EFS} is the reverse saturation current of the B-E diode and I_{CFS} that of the B-C diode, we get

$$I_E = I_{EFS} \left(\exp\left(\frac{V_{BE}}{V_T}\right) - 1 \right) - \alpha_R \, I_{CFS} \left(\exp\left(\frac{V_{BC}}{V_T}\right) - 1 \right)$$

$$(6.7)$$

$$I_C = - \, I_{CFS} \left(\exp\left(\frac{V_{BC}}{V_T}\right) - 1 \left(+ \alpha_F \, I_{EFS} \left(\exp\left(\frac{V_{BE}}{V_T}\right) - 1 \right) \right. \right.$$

Here all emission coefficients have been assumed to be unity. I_{EFS} and I_{CFS} correspond to the reverse saturation currents of the diodes obtained by shorting B-C and B-E junctions respectively.

Each of the diodes in Fig. 6.6(b) can be replaced by the model of Fig. 6.1 to get the complete Ebers-Moll model shown in Fig. 6.7. In this model the various parameters are as follows:

R_E, R_C : bulk and contact resistance of emitter and collector respectively (< 10 ohms)

R_B : base spreading resistance ($\leqslant 100$ ohms)

α_F : normal mode common base current gain

α_R : inverse mode common base current gain

R_{de} : B-E diode non-linearity given by

$$I_{EF} = I_{EFS} \left(\exp\left(\frac{V_{B'E'}}{\eta_E V_T}\right) - 1 \right)$$

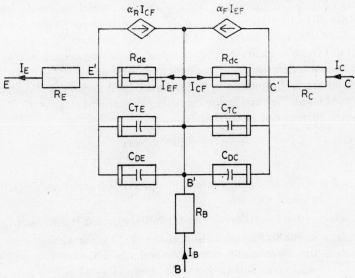

Fig. 6.7   AC Ebers-Moll model with the various non-linear capacitors included.

R_{de} : B–C diode non-linearity given by

$$I_{CF} = I_{CFS}\left(\exp\left(\frac{V_{B'C'}}{\eta_C V_T}\right) - 1\right)$$

(η_E and η_C) are emission coefficients)

C_{TE} : B–E junction transition capacitance given by

$$C_{TE} = \frac{C_{JE}}{\left(1 - \dfrac{V_{B'E'}}{V_{JE}}\right)^{m_E}}$$

where

V_{JE} = contact potential and m_E = grading coefficient.

C_{TC} : B–C junction transition capacitance given by

$$C_{TC} = \frac{C_{JC}}{\left(1 - \dfrac{V_{B'C'}}{V_{JC}}\right)^{m_C}}$$

C_{DE} : B–E junction diffusion capacitance given by

$$C_{DE} = \frac{I_{EFS}}{\eta_E V_T}\,\tau_F\,\exp\left(\frac{V_{B'E'}}{\eta_E V_T}\right)$$

C_{DC} : B–C junction diffusion capacitance given by

$$C_{DC} = \frac{I_{CFS}}{\eta_C V_T}\,\tau_R\,\exp\left(\frac{V_{B'C'}}{\eta_C V_T}\right)$$

The various expressions for the capacitances are similar to those for the diode given in Eqns. (6.6) and (6.3).

Though Fig. 6.7 represents a fairly elaborate model, it needs improvement. One of its limitations is that only one emission coefficient is provided for each of the two junctions. Even for the *pn* junction diode this was pointed out to be inaccurate. The transport model described below, gives a more accurate description of the emission constants.

Consider the $B-E$ junction to be forward biased and the $B-C$ junction to be reverse biased. Neglecting the currents originating in the $B-C$ junction, the emitter current can be split as follows:

$$i_E = i_{EB} + i_{BE} + i_{RTE}$$

where,

i_{EB} is due to carriers (electrons for NPN transistor) injected from emitter to base.

i_{BE} is due to carriers (holes for NPN) injected from base to emitter

i_{RTE} is due to recombination in the $B-E$ space charge region. Of the three components i_{EB} and i_{BE} will vary as exp (V_{BE}/V_T) while i_{RTE} will vary more like exp $(V_{BE}/2V_T)$. The base current also has three components

$$i_B = i_{BE} + i_{RTE} + i_{RNC}$$

where i_{RNC} is due to a recombination in the neutral base region which also, varies as exp (V_{BE}/V_T). Similarly, when the $B-C$ junction is forward biased we get another set of three components for the base current

$$i_B = i_{BC} + i_{RTC} + i_{RNC}$$

Here again, i_{BC} and i_{RNC} vary as exp (V_{BC}/V_T) while i_{RTC} varies roughly as exp $(V_{BC}/2V_T)$.

In other words, each of the junctions should have two non-linear resistors having exponential relations with $\eta = 1$ and $\eta = 2$ respectively. The two components of current i_{RTE} and i_{RTC} associated with space charge recombination do not contribute to currents transported across the base (i.e., from collector to emitter or vice versa). In the normal active mode

$$I_C \simeq \alpha_F I_E = \alpha_F I_{EFS} (\exp (V_{BE}/V_T) - 1)$$

and in the inverse active mode

$$I_E \simeq \alpha_R I_C = \alpha_R I_{CFS} (\exp (V_{BC}/V_T) - 1.)$$

In general, the current transported across the base can be written as the sum of the two and denoted by I_{CC}.

$$I_{CC} = \alpha_F I_{EFS} (\exp (V_{BE}/Y_T) - 1) - \alpha_R I_{CFS} (\exp (V_{BC}/V_T) - 1)$$

But $\alpha_N I_{EFS} = \alpha_R I_{CFS}$ [8].

Denoted $\alpha_N I_{EFS} = \alpha_R I_{CFS} = I_S$ we get,

$$I_{CC} = I_S [\exp (V_{BE}/V_T) - \exp (V_{BC}/V_T)]$$

Neglecting R_E, R_C and R_B, the DC transport model looks as shown in Fig. 6.8. The components of base current going to emitter and collector are

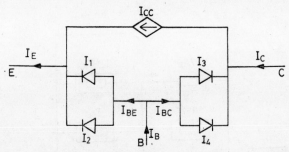

Fig. 6.8   DC transport Model for a BJT·I_1, I_2, I_3 and I_4 are the reverse saturation currents.

called I_{BE} and I_{BC}.

$$I_{BE} = I_1 \left(\exp \left(\frac{V_{BE}}{V_T} \right) - 1 \right) + I_2 \left(\exp \left(\frac{V_{BE}}{2T_T} \right) - 1 \right)$$

$$I_{BC} = I_3 \left(\exp \left(\frac{V_{BC}}{V_T} \right) - 1 \right) + I_4 \left(\exp \left(\frac{V_{BC}}{2V_T} \right) - 1 \right)$$

For large forward bias V_{BE} and reverse bias V_{BC},

$$I_B \simeq I_1 \ \exp \left(\frac{V_{BE}}{V_T} \right)$$

and

$$I_C \simeq I_S \ \exp \left(\frac{V_{BE}}{V_T} \right)$$

The ratio I_C/I_B must correspond to B_F and, therefore

$$I_1 = \frac{I_S}{\beta_F} = \frac{I_S(1 - \alpha_F)}{\alpha_F}$$

Similarly, $I_3 = \dfrac{I_S}{B_R}$. In the SPICE model I_2 and I_4 are called I_{SE} and I_{SC} respectively. Now we can write,

$$I_{BE} = \frac{I_S}{\beta_F} \left(\exp \left(\frac{V_{BE}}{V_T} \right) - 1 \right) + I_{SE} \left(\exp \left(\frac{V_{BE}}{2V_T} \right) - 1 \right)$$

$$I_{BC} = \frac{I_S}{\beta_R} \left(\exp \left(\frac{V_{BC}}{V_T} \right) - 1 \right) + I_{SC} \left(\exp \left(\frac{V_{BC}}{2V_T} \right) - 1 \right)$$

The various capacitances remain the same as in the Ebers-Moll model. The complete transport model can be represented as shown in Fig. 6.9.

The actual model used in the SPICE program is the Gummel-Poon model [9]. It takes care of the following additional features of the transistor.

1. When the B–E junction is so highly forward biased that the base minority carrier concentration approaches the majority carrier concentration, high injection is said to occur. Under such conditions the collector current varies as $\exp(qV_{BE}/2kT)$ rather than as $\exp(qV_{BE}/kT)$. This is accounted for in the Gummel-Poon model.

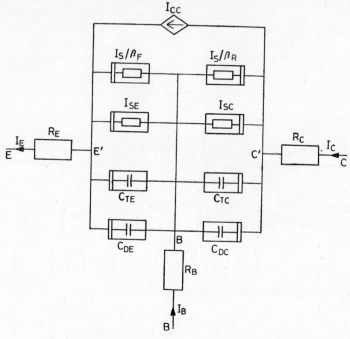

Fig. 6.9   AC transport model with capacitors included

2. For a constant base current, I_c varies as a function of V_{BE}. This corresponds to the slope of the characteristics in the common emitter configuration. This effect, known as Early effect is also accounted for.

Both these effects are catered to by a term Q_B in the definition of I_{CC}. I_{CC} is written as

$$I_{CC} = \frac{I_S}{Q_B}\left(\exp\left(\frac{V_{BE}}{V_T}\right) - \exp\left(\frac{V_{BC}}{V_T}\right)\right)$$

The term Q_B stands for the net charge in the base associated with majority carriers (positive for *npn* transistors). It is a normalised quantity, the charge for zero bias being taken as unity. Q_B is defined in terms of two other quantities Q_1 and Q_2 as

$$Q_B = \frac{Q_1}{2}\left[1 + (1 + 4Q_2)^{1/2}\right]$$

where

$$Q_1 = \frac{1}{1 - \dfrac{V_{BC}}{V_{AF}} - \dfrac{V_{BE}}{V_{AR}}}$$

$$Q_2 = \frac{I_S}{I_{KF}}\left(\exp\left(\frac{V_{BE}}{V_T}\right) - 1\right) - \frac{I_S}{I_{KR}}\left(\exp\left(\frac{V_{BC}}{V_T}\right) - 1\right)$$

The quantity Q_1 caters to the Early effect and Q_2 to high injection effects. Let us consider these one by one. If $Q_2 = 0$ (no high injection),

$$Q_B = Q_1 = \frac{1}{1 - \dfrac{V_{BC}}{V_{AF}} - \dfrac{V_{BE}}{V_{AR}}}$$

Figure 6.10(a) shows the common emitter characteristic for a BJT. The various lines at constant I_B when extended in the negative V_{CE} direction roughly meet at a point. The voltage (in magnitude) here is termed the forward Early voltage V_{AF}. It is usually a large like 200 V. Figure 6.10(b) shows V_{AR} similarly defined as the reverse Early voltage in the inverse

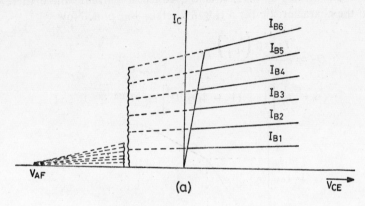

(a)

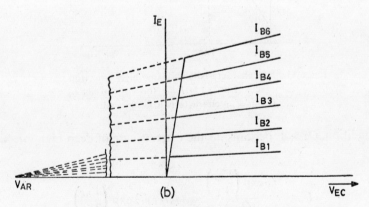

(b)

Fig. 6.10   BJT characteristics showing (a) Forward Early voltage;
(b) Reverse Early voltage.

active mode. In the normal active mode V_{BC} is negative and of the order of a few volts while V_{BE} is positive and about 0.7 V. Therefore,

$$Q_1 \simeq \frac{1}{1 + \dfrac{|V_{BC}|}{V_{AF}}} \simeq \left(1 - \frac{|V_{BC}|}{V_{AF}}\right)$$

if $| V_{BC} | \ll V_{AF}$. The current I_{CC} and therefore I_C increases with $| V_{BC} |$. As $V_{CE} \simeq | V_{BC} | + 0.7$ V, I_C increases with V_{CE} as well. It may be mentioned in passing, that the Early effect is due to base width modulation. For larger reverse bias of the B–C junction the depletion region extends further into the base. This reduces the possibility of recombination in the neutral base region thereby increasing α and the ratio I_C/I_B. This effect is modelled by the change in the charge of the space charge region Q_1.

The charge Q_2 takes care of high injection. Once again consider the normal active region. Let $Q_1 = 1$, i.e. early effect is neglected. High Injection takes place when I_C approaches I_{KF}. I_{KF} is defined as the knee current as shown in Fig. 6.11. It is the collector current where there is a knee in the characteristic for a (log I_C) versus V_{BE} plot. Now

$$Q_2 = \frac{I_S \exp \left(\dfrac{V_{BE}}{V_T}\right)}{I_{KF}}$$

$$Q_B = \frac{1}{2}\left[1 + (1 + 4Q_2)^{1/2} \right] \simeq Q_2^{1/2} \text{ if } Q_2 \gg 1$$

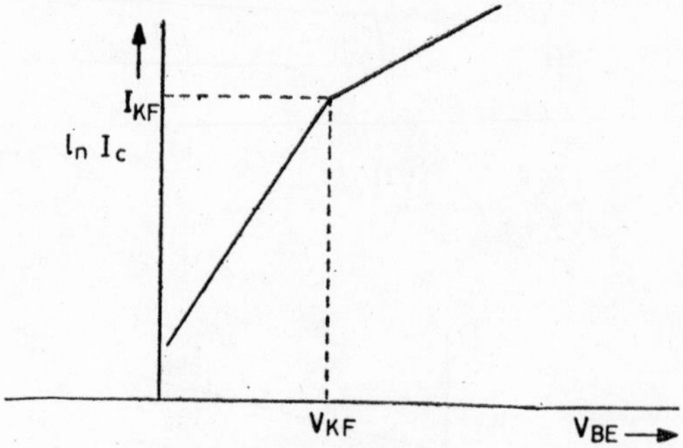

Fig. 6.11   BJT characteristic showing high injection effects. For $I_C \gg I_{KF}$, the slope of log I_C versus V_{BE} characteristic is reduced.

The condition $Q_2 \gg 1$ pertains to the extreme case of deep high injection. Then

$$I_C = \frac{I_S \exp \left(\dfrac{V_{BE}}{V_T}\right)}{Q_B} = (I_S I_{KF})^{1/2} \exp \left(\frac{V_{BE}}{2V_T}\right)$$

We do indeed get a variation like $\exp (V_{BE}/2V_T)$ for high injection. The term I_{KR} corresponds to the knee in the characteristic when (log I_E) is plotted versus V_{BC} in the inverse active mode.

We have written the above equations assuming $\eta = 1$ for the I_{CC}, I_S/B_F and I_S/B_R components of the current and $\eta = 2$ for I_{SE} and I_{SC} components. The SPICE model allows the user to specify a η_F for the V_{BE} exponent of I_{CC} and I_S/B_F, a η_R for the V_{BC} exponent of I_{CC} and I_S/B_R, a η_C for I_{SC} and a η_E for I_{SE}. The equations for the collector and base currents in the DC SPICE Gummel-Poon model are

$$I_C = \frac{I_S}{Q_B}\left(\exp\left(\frac{V_{BE}}{\eta_F V_T}\right) - \exp\left(\frac{V_{BC}}{\eta_R V_T}\right)\right) - \frac{I_S}{\beta_R}\left(\exp\left(\frac{V_{BC}}{\eta_F V_T}\right) - 1\right)$$

$$- I_{SC}\left(\exp\left(\frac{V_{BC}}{\eta_C V_T}\right) - 1\right)$$

$$I_B = \frac{I_S}{\beta_F}\left(\exp\left(\frac{V_{BE}}{\eta_F V_T}\right) - 1\right) + \frac{I_S}{\beta_R}\left(\exp\left(\frac{V_{BC}}{\eta_R V_T}\right) - 1\right)$$

$$+ I_{SE}\left(\exp\left(\frac{V_{BE}}{\eta_E V_T}\right) - 1\right) + I_{SE}\left(\exp\left(\frac{V_{BC}}{\eta_C V_T}\right) - 1\right)$$

The SPICE Gummel-Poon model is slightly different from the actual Gummel-Poon model [9]. We have, all along, considered only the SPICE Gummel-Poon model. Some of the other parameters of the SPICE BJT model are:

IRB, RBM: Apart from the zero bias base resistance *RB, IRB* and *RBM* are used to model the decrease of base resistance at high currents because of emitter crowding [10]. *IRB* is the value of I_C at which base resistance falls halfway to its minimum value *RBM* from *RB* as shown in Fig. 6.12.

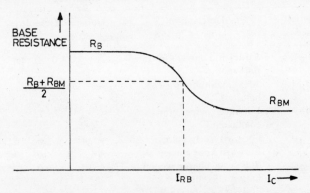

Fig. 6.12   Variation of base resistance with collector current

XTF, VTF, ITF: The transit time τ_F (used to model the *B–E* diffusion capacitance) varies with bias. These three parameters model this effect. For details the reader is referred to the SPICE manual [6].

PTF: The common emitter unity current gain frequency or α cut off frequency is given by $f_T = 1/2\pi\tau_F$. The effect of higher order poles would be to change the phase at f_T. This excess phase is given by *PTF*. If *PTF* = 0, the phase would be 90°.

CJS, MJS, VJS: These are used to describe the depletion capacitance C_{TS} between the collector and substrate. Like other depletion capacitances it is given by

$$C_{TS} = \frac{CJS}{\left(1 - \frac{V}{VJS}\right)^{MJS}}$$

XCJC: This gives the fraction of *B–C* transition capacitance connected to internal base node B' as opposed to B. The default value is 1, i.e., all the *B–C* depletion capacitance is connected to B' as shown in Fig. 6.9.

XTB: This gives the temperature variation of β_R, β_F, I_{SE} and I_{SC}

$$\beta_F = \beta_{FO}\left(\frac{T}{T_0}\right)^{XTB}, \ \beta_R = \beta_{RO}\left(\frac{T}{T_0}\right)^{XTB}$$

$$I_{SE} = I_{SEO}\left(\frac{T}{T_0}\right)^{XTI-XTB} e^{qEG(T-T_0)/kTT_0}$$

$$I_{SC} = I_{SCO}\left(\frac{T}{T_0}\right)^{XTI-XTB} e^{qEG(T-T_0)/kTT_0}$$

(*EG* is in *Vs* and k in joules/°K).

Quantities *KF* and *AF* describe flicker noise and are described in Sec. 6. *EG* and *XTI* describe the temperature variation of I_S as for the diode. *FC* is the coefficient to modify all depletion capacitances when the forward bias voltage approaches the contact potential of the junction. Table 6.2 below gives a list of SPICE BJT parameters.

Table 6.2:   List of SPICE BJT Parameters

Name	Symbol used	Description	Units	Default	Typical
IS*	I_S	Saturation current in transport model	A	1.0×10^{-16}	1.0×10^{-15}
BF	β_F	Ideal maximum forward beta	—	100	100
NF	η_F	Forward current emission coefficient	—	1	1
VAF	V_{AF}	Forward Early voltage	V	∞	200
IKF*	I_{KF}	Forward knee current	A	∞	0.01
ISE*	I_{SE}	B–E space charge recombination current	A	0	1.0×10^{-13}
NE	η_E	Emission coefficient for I_{SE}	—	1.5	2.0
BR	β_R	Ideal maximum reverse beta	—	1.0	0.1
NR	η_R	Reverse current emission coefficient	—	1.0	1.0
VAR	V_{AR}	Reverse Early voltage	V	∞	200
IKR*	I_{KR}	Reverse knee current	A	∞	0.01
ISC*	I_{SC}	B–C space charge recombination current	A	0	1.0×10^{-13}
NC	η_C	Emission coefficient for ISC	—	2	1.5
RB*	RB	Zero bias base resistance	Ohms	0	100
IRB	IRB	Current where base resistance falls halfway to its minimum value	A	∞	0.1
RBM*	RBM	Minimum base resistance at high current	Ohms	RB	10

Name	Symbol used	Description	Units	Default	Typical
RE*	R_E	Emitter resistance	Ohms	0	1
RC*	R_C	Collector resistance	Ohms	0	10
CJE*	C_{JE}	B–E zero bias depletion capacitance	F	0	2 pF
MJE	M_E	B–E junction grading coefficient	—	0.33	0.33
VJE	V_{JE}	B–E built-in or contact potential	V	0.75	0.6
TF	τ_F	Ideal forward transit time	sec	0	0.1 nsec
XTE	XTF	Coefficient for bias dependence of TF	—	0	
VTF	VTF	Voltage describing V_{BC} dependence of TF	V	∞	
ITF*	ITF	High current parameters for effect on TF	A	0	
PTF	PTF	Excess phase at $f = 1.0/2\pi TF$	deg	0	
CJC*	C_{JC}	B–C zero-bias depletion capacitance	F	0	2 pF
VJC	V_{JC}	B–C built-in or contact potential	V	0.75	0.5
MJC	M_C	B–C junction grading coefficient	—	0.33	0.5
XCJC	$XCJC$	Fraction of B–C depletion capacitance connected to internal base note	—	1.0	
TR	τ_R	Ideal reverse transit time	sec	0	10 nsec
CJS*	CJS	Zero-bias collector-sub-strate depletion capacitance	F	0	2 pF
VJS	VJS	Collector-substrate junction built-in potential	V	0.75	
MJS	MJS	Substrate junction grading coefficient	—	0	0.5
XTB	XTB	Forward and reverse beta temperature exponent	—	0	
EG	EG	Energy gap for temperature effect on IS	eV	1.11	1.11
XTI	XTI	Temperature exponent for effect on IS	—	1.3	3
KF	K_f	Flicker noise coefficient	—	0	
AF	A_f	Flicker noise exponent	—	1	
FC	FC	Coefficient for forward bias depletion capacitance formula	—	0.5	

Parameters marked with an asterisk are area dependent

A BJT, being an highly complex device, can never be fully described by any model. The SPICE model also, naturally, has some limitations. The description is really best suited only for the normal active region and perhaps saturation region. Many assumptions of the Gummel-Poon model itself breakdown in the inverse active region. The various breakdown mechanisms in a transistor are also not included. As in the case of the diode model for reverse bias, actual transistor currents in cutoff will be much larger than those given by the model. It is well known that the collector or base current I_{CBO}, with the emitter open and C–B junction reverse biased can be large and unpredictable in a real transistor. This is due to surface leakage, space charge generation and many other second order effects. The SPICE model cannot account for these effects. This is not so much a limitation as a manifestation of the fact that some compromise has to be struck between accuracy and computational complexity.

6.4 Model for MOSFETS

Most LSIs/VLSIs made today are based on MOS (metal oxide semi-conductor) technology. Larger packing densities are possible with MOS. Common MOS technologies today are NMOS and CMOS. While NMOS is faster, CMOS has very little DC power dissipation, excellent noise margin and can function with a wide range of supply voltages. With advances in technology CMOS logic circuits are getting faster with speeds now approaching NMOS and even TTL. Linear ICs like opamps are also increasingly being made with CMOS technology. It is quite clear that a circuit model for a MOSFET (metal oxide semiconductor field effect transistor) should be built into any circuit analysis program. The model described here is with reference to the model used in SPICE [6]. NMOS transistors are assumed. Models are easily extended to PMOS with appropriate polarity changes. We first discuss the DC model and later on introduce the capacitances.

6.4.1 DC Model

Figure 6.13 shows the basic structure of a NMOS transistor. There is a p type substrate on which there are two n^+ wells. These two wells constitute

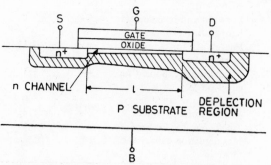

Fig. 6.13 Typical structure of a NMOS transistor

the source and drain. In between and above the wells is a thin silicon dioxide layer and over it the gate. The gate can be metal or more commonly heavily doped polycrystalline silicon. The term metal in MOS has remained though there is usually no metal in the gate of MOSFETS today. The substrate in a NMOS IC is common to all transistors and connected to the lowest voltage. A transistor is called NMOS or PMOS depending on whether a n channel or p channel forms below the oxide between the source and drain. For a MOS transistor, the channel as well as the source and drain are separated from the substrate by the depletion region of a reverse biased pn junction diode. The gate draws virtually no DC current and the input impedance at the gate is mainly capacitive.

Let the source and drain terminals be open and starting from a negative value let increasing voltages be applied to the gate. The substrate is grounded. Three stages can be identified as a n channel ultimately forms below the oxide (for a NMOS device).

(a) *Accumulation layer*: For negative gate voltages, holes in the substrate are attracted to the oxide substrate interface. A p channel with increased hole density forms just below the oxide. This is known as an accumulation layer (Fig. 6.14(a)).

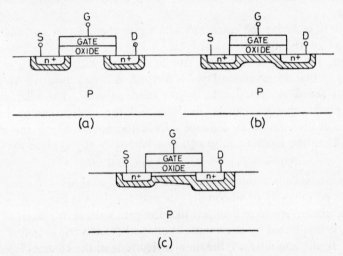

Fig. 6.14   Structure of a NMOS transistor showing (a) Accumulation;
(b) Depletion; (c) Inversion

(b) *Depletion layer*: As the gate voltage is increased holes are repelled from the oxide substrate interface. The region below the oxide becomes free of all mobile charges and becomes a space charge or depletion region. Only immobile acceptor ions are present (Fig. 6.14(b)).

(c) *Inversion layer*: For even larger gate voltages mobile electrons are attracted from the deeper substrate to the oxide substrate interface. These mobile electrons form a n channel. If a voltage is applied between source and drain, electrons will flow towards the positive terminal. A transistor which is ON has a inversion layer below the oxide and an OFF transistor a depletion or accumulation layer. Even when an inversion layer exists there will be a depletion region separating the channel or inversoin layer from the substrate (Fig. 6.14(c)).

In a NMOS enhancement mode transistor a positive voltage must be applied to the gate (relative to substrate) before a n channel forms and conduction takes place. In a depletion mode transistor an inversion layer exists even for small negative voltage. A sufficiently large negative voltage must be applied to turn off the device. The gate voltage at which the inversion layer forms and has a electron density equal to the hole density deep in the substrate is called the threshold voltage V_T. When this happens strong inversion is said to take place. V_T is positive for an enhancement mode transistor, and negative for a depletion mode transistor. One may wonder how, in a depletion mode transistor, an inversion layer exists for negative gate voltages. The description given above is simplistic. The oxide substrate interface is very complex and there are positive immobile charges near the interface in the oxide. These charges tend to make the threshold voltage negative. In any case for proper control, a thin layer of donor ions is usually ion implanted in the substrate below the oxide to make a depletion transistor.

We will now derive the equations which give the channel current I_{DS} as a function of the four terminal voltages V_B, V_S, V_D and V_G (B stands for substrate). First we consider the simple situation with $V_S = V_B = 0$. Let $V_G > V_T$ so that an inversion layer is formed. A small positive voltage is applied at the drain relative to source (and ground). The relation for a parallel plate capacitor can be used to find the mobile charge in the n channel. Let the voltage at distance y from the source along the channel be $V(y)$. Then the mobile charge dQ_I in a length dy at y is given by

$$dQ_I = -[V_G - V_T - V(y)]C_{OX}W \, dy \qquad (6.8)$$

$C_{OX} =$ capacitance per unit area between gate and channel $= \epsilon_{OX}/t_{OX}$ where ϵ_{OX} is the permittivity of silicon dioxide and t_{OX} the thickness, $W =$ width of the transistor. We have assumed that the gate voltage for onset of inversion, V_T, is unaffected by the nonzero voltage $V(y)$ at various points on the channel. If the mobility of the mobile electrons in the channel is μ_n, then the drain to source current I_{DS} is given by

$$I_{DS} = \mu_n \frac{dQ_T}{dy} \frac{dV}{dy} = \mu_n C_{OX} W [V_G - V_T - V(y)] \frac{dV}{dy}$$

Integrating we get

$$\int_0^l I_{DS} \, dy = \mu_n C_{OX} W \int_{V_S}^{V_D} (V_G - V_T - V[(y)) \, dV$$

For $V_S = 0$

$$I_{DS} = \frac{\mu_n C_{OX} W}{l} \left[(V_{GS} - V_T) V_{DS} - \frac{V_{DS}^2}{2} \right]$$

The quantity $\frac{\mu_n C_{OX} W}{l}$ is called the transconductance parameter $\beta(KP$ in SPICE) and has the units A/V^2.

As V_{DS} is increased (keeping $V_S = V_B = 0$), the part of the channel near the drain has a larger voltage. The voltage across the oxide is therefore less. Finally, when $V_{DS} = V_{GS} - V_T$, there is no longer an inversion region near the drain but a depletion region as shown in Fig. 6.15. The n channel ends some distance from the drain and electrons have to cross the depletion region to reach the drain. The electric field in the y direction has the right polarity to aid this flow. Any increase in V_{DS} beyond $(V_{GS} - V_T)$ only increase the length of the depletion region with I_{DS} remaining roughly constant. The extra voltage $[V_{DS} - (V_{GS} - V_T)]$ appears across the depletion region. The current I_{DS} in this saturation region is given by substituting $V_{DS} = (V_{GS} - V_T)$ in the above expression for I_{DS}. This gives

$$I_{DS\text{sat}} = \frac{\beta}{2}(V_{GS} - V_T)^2$$

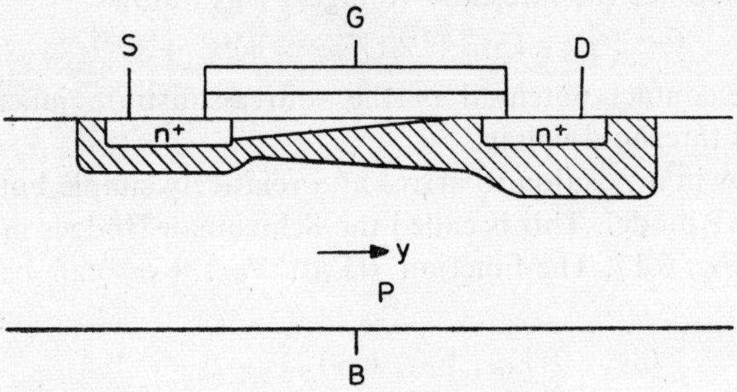

Fig. 6.15   Saturation in an NMOS transistor

The I_{DS} versus characteristics can now be sketched as shown in Fig. 6.16. The dotted line is defined by $V_{DS} = V_{GS} - V_T$ and separates the saturation

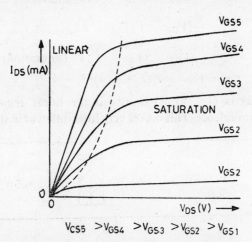

Fig. 6.16   Typical $I_{DS} - V_{DS}$ characteristics for a NMOS transistor. The line $V_{DS} = V_{GS} - V_T$ demarcates the linear and saturation regions.

region from the linear or triode region. Well to the left of the line, I_{DS} varies almost linearly with V_{DS}. To the right I_{DS} is roughly constant. Even in the saturation region, I_{DS} increases slightly as V_{DS} increases (for constant V_{GS}). This is because the effective length and therefore resistance of the channel is reduced as V_{DS} is increased. An empirical relation is often used to describe the variation of I_{DS} in the saturation region. A channel length modulation parameter λ is used and one writes,

$$I_{DSsat} = \frac{\beta}{2}(V_{GS} - V_T)^2(1 + \lambda V_{DS})$$

where λ is typically less than 0.1.

In case $V_S \neq 0$ and there is a finite source to substrate bias V_{SB} (note V_{SB} must be positive for NMOS), this is taken care of in the simple model described so far by a factor γ. This factor γ is called the bulk threshold factor and modifies the threshold voltage V_T as follows

$$V_T = V_{TO} + \gamma[(V_{SB} + \phi)^{1/2} - \phi^{1/2}]$$

Here ϕ is the contact potential of the source-substrate junction and V_{TO} the zero bias threshold voltage.

We are now in a position to arrive at a relatively simple but very popular DC MOSFET model. This is called the Schichman-Hodges model [11] and is shown in Fig. 6.17. The function $f(V_{DS}, V_{GS}, V_{SB})$ can be described as follows.

$$I_{DS} = f(V_{GS}, V_{DS}, V_{SB})$$

$$f(V_{GS}, V_{DS}, V_{SB}) = 0 \quad \text{for } V_{GS} < V_T \text{ (cut off)}$$

$$= \beta\left[(V_{GS} - V_T)V_{DS} - \frac{V_{DS}^2}{2}\right] \quad \text{for } V_{GS} > V_T$$

$$V_{DS} < (V_{GS} - V_T) \text{ (linear)}$$

$$= \frac{\beta}{2}(V_{GS} - V_T)^2(1 + \lambda V_{DS}) \text{ for } V_{GS} > V_T$$

$$V_{DS} \geqslant (V_{GS} - V_T) \quad \text{(saturation)}$$

where $\qquad V_T = V_{TO} + \gamma[(V_{SB} + \phi)^{1/2} - \phi^{1/2}]$ \hfill (6.9)

Sometimes the factor $(1 + \lambda V_{DS})$ is used in the linear region as well for programming convenience. This makes very little different in the linear region.

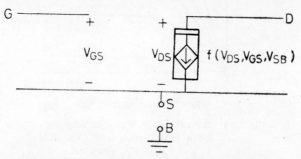

Fig. 6.17 Simple DC non-linear MOSFET model

In Eq. (6.8) for dQ_I we assumed V_T to be independent of $V(y)$. The threshold voltage V_T, can be more accurately written as ([12], [13])

$$V_T = \frac{-Q_{BO}}{C_{OX}} - \frac{Q_{OX}}{C_{OX}} + \phi_{GC} + 2\phi_F \qquad (6.10)$$

where

$Q_{BO} = -[2\epsilon_{Si}qN_A(2\phi_F + V(y))]^{1/2}$

ϵ_{Si} = permittivity of silicon

N_A = acceptor concentration in substrate

ϕ_F = Fermi potential in substrate

Q_{OX} = Charge in the region near the substrate-oxide interface. This is inside the oxide, immobile and positive for NMOS and PMOS

ϕ_{GC} = difference in Fermi potential between gate and substrate.

Q_{BO} represents the charge due to the exposed acceptor ions in the channel just before inversion occurs. Its variation along y is what makes our earlier analysis inaccurate. The origin of Q_{OX} is quite complex and partly due to surface state and partly due to trapped oxide charge [14]. The term $2\phi_F$ reflects the fact that strong inversion is said to take place when the channel electron density equals that of the holes deep in the substrate. This implies the bands have to be bent by $2\phi_F$. For a NMOS transistor two of the terms listed in Eq. (6.10) are positive ($-Q_{BO}/C_{OX}$ and $2\phi_F$) and two of the them negative (ϕ_{GC} and $-Q_{OX}/C_{OX}$). As an example consider a polysilicon gate MOSFET with the following specifications

N_A (of substrate) $= 5 \times 10^{14}/\text{cm}^3 = 5 \times 10^{20}/\text{m}^3$

N_D (of gate) $\qquad = 10^{20}/\text{cm}^3 \qquad = 10^{26}/\text{m}^3$

$t_{OX} \qquad\qquad = 330 \times 10^{-10}$ metres

$Q_{OX} \qquad\qquad = 10^{10}/\text{cm}^2 \qquad = 10^{14}/\text{m}^2$

V_{TO} (for $V_{SB} = 0$) can be calculated term by term from Eq. (6.10)

$$\phi_F = \frac{-kT}{q} \ln\left(\frac{n_i}{N_A}\right) = -0.026 \ln\left(\frac{1.45 \times 10^{10}}{5 \times 10^{14}}\right) = 0.27 \text{ V}$$

$$\phi_{GC} = \phi_{FG} - \phi_{F\,\text{sub}} = \frac{-kT}{q} \ln\left(\frac{10^{20}}{1.45 \times 10^{10}}\right) -0.27 = -0.85 \text{ V}$$

$$Q_{BO} = -(2 \times 12 \times 8.854 \times 10^{-12} \times 1.6 \times 10^{-19} \times 5 \times 10^{20} \times 0.54)^{1/2}$$

$$= -9.58 \times 10^{-5} \text{ coulombs/m}^2$$

$$C_{OX} = \frac{\epsilon_{OX}}{t_{OX}} = \frac{3.9 \times 8.854 \times 10^{-12}}{330 \times 10^{-10}} = 1.046 \times 10^{-3} \text{ F/m}^2$$

$$V_{TO} = \frac{9.58 \times 10^{-5} - 1.6 \times 10^{-19} \times 10^{14}}{1.046 \times 10^{-3}} + 0.54 - 0.85$$

$$= -0.234 \text{ V}$$

Table 6.3 gives the sign for various terms in Eq. (6.10) for NMOS and PMOS transistors.

Table 6.3

Parameter	NMOS	PMOS	Typical values (magnitude)
ϕ_F	+	−	0.3 V
Q_{BO}	−	+	−
Q_{OX}	+	+	$10^{10}/\text{cm}^2$
ϕ_{GC}			
(a) n^+ gate	−	−	0.9 V
(b) p^+ gate	+	+	0.3 V
(c) Al gate	−	−	0.3 V

The sum of the terms ϕ_{GC} and $(-Q_{OX}/C_{OX})$ is termed the flat band voltage V_{FB}. It is the voltage needed at the gate to make the region just below the oxide similar to that deep in the substrate i.e. the gate voltage needed to keep the band flat

$$V_T = \frac{-Q_{BO}}{C_{OX}} + V_{FB} + 2\phi_F$$

Using the above expression for V_T and expressing Q_{BO} as a function of $V(y)$ we can arrive at a more accurate expression for I_{DS} as

$$I_{DS} = \frac{\mu_n C_{OX} W}{l} \int_{V_S}^{V_D} [V_G - \{\gamma(2\phi_F + V(y))^{1/2} + V_{FB}$$
$$+ 2\phi_F\} - V(y)]\, dV$$

where $\gamma = (2\epsilon_{si}qN_A)^{1/2}/C_{OX}$ and all voltages are with respect to the substrate. If $V_{SB} \neq 0$, let $V'(y) = V(y) - V_{SB}$. Then, $V_G = V_{GS} + V_{SB}$, $dV = dV'$ and $V(y) = V'(y) + V_{SB}$

$$I_{DS} = \beta \int_0^{V_{DS}} [(V_{GS} - V_{FB} - 2\phi_F) - \gamma(2\phi_F + V(y))^{1/2} - V'(y)]\, dV'$$

$$= \beta \left[V_{DS}\left(V_{GS} - V_{FB} - 2\phi_F - \frac{V_{DS}}{2}\right) - \frac{2}{3}\gamma\{(V_{DS} + 2\phi_F \right.$$

$$\left. + V_{SB})^{1/2} - (2\phi_F + V_{SB})^{3/2}\} \right] \qquad (6.11a)$$

The above equation applies in the linear region. The drain to source voltage for onset of saturation is found by putting $\dfrac{dI_{DS}}{dV_{DS}} = 0$. This gives

$$V_{DS\,sat} = V_{GS} - V_{FB} - 2\phi_F + \gamma^2 - \gamma\left[\frac{\gamma^2}{4} + V_{GS} - V_{FB} + V_{SB}\right]^{1/2}$$

$$\qquad (6.11b)$$

and

$$I_{DS\,sat} = \beta/2\bigg\{(V_{GS} - V_{FB} - 2\phi_F)^2 - F^2(\gamma, V_{GS}, V_{SB})$$

$$-\frac{4}{3}\,\gamma\{[V_{GS} - V_{FB} + F(\gamma, V_{GS}, V_{SB})]^{3/2} - (2\phi_F + V_{SB})^{3/2}\}\bigg\}$$

(6.11c)

where,

$$F(\gamma, V_{GS}, V_{SB}) = \frac{\gamma^2}{2} - \gamma\left(V_{GS} - V_{FB} + \frac{\gamma^2}{4} + V_{SB}\right)$$

The variation of V_T with source to substrate voltage is still given by Eq. (6.9) where γ is the bulk threshold factor and equal to $(2\epsilon_{Si}qN_A)^{1/2}/C_{OX}$.

The slight increase in I_{DS} with V_{DS} can be modelled more accurately than in Eq. (6.9) by modifying the length. We write

$$l' = l - \Delta l$$

where Δl is the width of the depletion region between channel and drain. From the relation for a *pn* junction diode (Eq. 6.2)

$$\Delta l = \frac{2\epsilon_{Si}}{qN_A}(V_{DS} - V_{DS\,sat})^{1/2}$$

The above expression overestimates the increase in current partly because velocity saturation is neglected and partly because the electrons in the depletion region make the charge density greater than N_A [12]. A more accurate expression for Δl is given by Baum [15] and used in SPICE. Assuming the source end of the depletion region is at $V_{DS\,sat}$,

$$\Delta l = \left[\left(\frac{E_{max}}{2a}\right)^2 + \frac{V_{DS} - V_{DS\,sat}}{a}\right]^{1/2} - \frac{E_{max}}{2a}$$

where $a = \frac{qN_A}{2\epsilon_{Si}}$ and E_{max} is the critical lateral electric field at which velocity saturation occurs [see Fig. 6.18]. A couple of empirical factors, NEFF and KAPPA, are further used to improve the accuracy. $(V_{DS} - V_{DS\,sat})$ is written as KAPPA $(V_{DS} - V_{DS\,sat})$ and N_A as (NEFF)N_A.

Another important effect to be modelled is the variation of mobility μ_n. It is first a function of temperature. Then it is a function of the vertical and lateral electric fields in the channel. It is written as

$$\mu_n = \mu_{n0}(T)f_v(V_G, V_S, V_D)\,f_h(V_G, V_S, V_D)$$

where f_v gives the dependence on vertical electric field and f_h the dependence on horizontal electric field. As far as temperature variation is concerned, the mobility μ_{n0} at temperature T is found from that at temperature T_0 by [12]

$$\mu_{n0}(T) = \mu_{n0}(T_0)\left(\frac{T}{T_0}\right)^{-M}$$

where T and T_0 are in °K and M is an empirical constant with value 3 in SPICE. The function f_v is modelled in two ways in SPICE. In the LEVEL 3 model f_v is a function of an empirical parameter θ and given by

$$f_v = \frac{1}{1 + \theta(V_{GS} - V_T)}$$

In the LEVEL 2 model it is written as

$$f_v = \left(\frac{V_C}{V_{\text{avg}}}\right)^U$$

V_C = critical voltage = $E_{\text{crit}}t_{OX}$ and E_{crit} is the vertical field at which mobility begins to reduce.

$E_{\text{crit}} \simeq 10^5$ V/m

V_{avg} = average voltage across channel

$$= V_{GS} - V_T - \frac{V_{DS}}{2}$$

U = exponent whose value is about 0.25 for NMOS and 0.15 for PMOS ([12]), [17].

The function f_h depicts the fact that the velocity of carriers does not increase linearly with electric field at values beyond about 10^4 V/cm as shown in Fig. 6.18. The saturation velocity [17] (about 10^7 cm/s) is the same for PMOS and NMOS. Speed advantages enjoyed by NMOS over CMOS and PMOS are no longer valid. The saturation velocity decreases with temperature [18]. If s_{max} is the saturation velocity then f_h is written as

$$f_h = \frac{1}{1 + \mu_{n0}f_t\left(\dfrac{V_{DS}}{l}\right)\dfrac{1}{s_{\text{max}}}} \quad \text{for } V_{DS} < V_{DS\,\text{sat}}$$

$$= \frac{1}{1 + \mu_{n0}f_v\left(\dfrac{V_{DS\,\text{sat}}}{l}\right)\dfrac{1}{s_{\text{max}}}} \quad \text{for } V_{DS} \geqslant V_{DS\,\text{sat}}$$

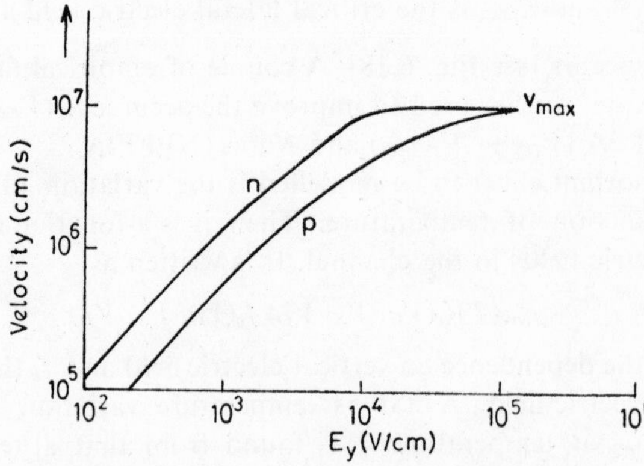

Fig. 6.18   Variation of drift velocity with electric field for n and p type materials. Note that the saturation velocities are not very different (from [15]).

6.4.2 AC or Capacitance Model

The capacitance model for a MOSFET is rather complex. There are four terminals and six possible capacitances. It has even been proposed that these are not reciprocal [19] ($C_{GS} \neq C_{SG}$) which then makes for 12 possible capacitances. More correctly, if one terminal is grounded (say substrate), then we have a (3×3) capacitance matrix with diagonal terms representing capacitances with respect to ground. The discussion here is restricted to incremental capacitances of the form $\partial Q / \partial V$. Generally, one tries to find the charge as a function of a various voltages and the capacitances are arrived at by differentiating with respect to the voltages. The reciprocal capacitance model in SPICE is based on the Meyer model [20] and the non-reciprocal model on the Ward and Dutton model [19]. We first discuss the simpler Meyer model. It is best to consider 3 different cases corresponding to accumulation, depletion and inversion layers below the oxide (see Fig. 6.14).

(a) *Accumulation layer*:   The capacitance C_{GB} is just C_{ox} as the substrate extends upto the oxide. C_{SB} and C_{DB} are just the depletion capacitances between substrate and source and drain respectively. C_{GS} and C_{GD} are the 'overhang' capacitances. They are present because the gate extends over the source and drain regions on either side. Figure 6.19 shows the various relevant capacitance for an accumulation layer in the channel region.

(b) *Depletion layer*:   Now C_{GB} is the series combination of C_{ox} and the depletion capacitance of the region below the oxide. Therefore C_{GB} is considerably less than C_{ox}. There is a continuous depletion region between drain and source. C_{DS} is just this depletion capacitance. C_{SB} and C_{DB} are again just the corresponding depletion capacitances. C_{GD} and C_{GS} are the overhang capacitances again. Various relevant capacitances are shown in Fig. 6.20.

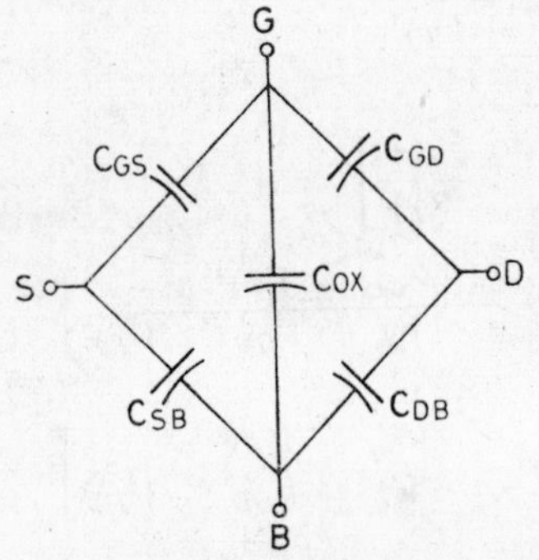

Fig. 6.19   MOSFET capacitances with an accumulation region

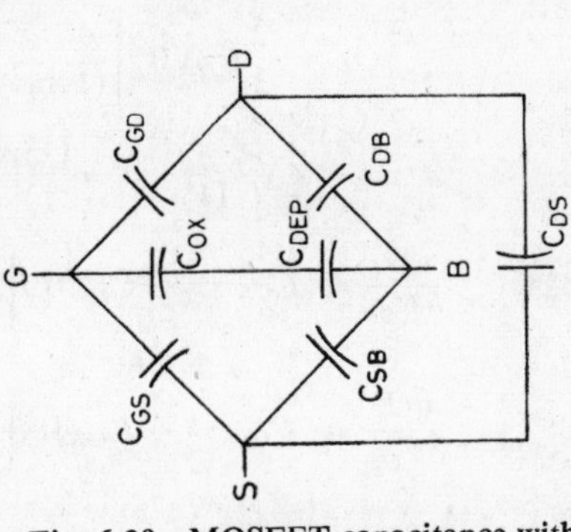

Fig. 6.20   MOSFET capacitance with a depletion region

(a) *Inversion layer*:   The situation now is far more complex. C_{SB} and C_{DB} are still the depletion capacitances of the reverse biased source-substrate and drain-substrate *pn* junctions. The gate is now not directly and capacitively linked to the substrate because of the channel. However, C_{GD} and C_{GS} now have two components

(a) Overhang capacitance
(b) Channel charge capacitance

The channel charge capacitance is a simplification of a transmission line model [23]. The transmission line model is replaced by two capacitances—one C_{GS} and the other C_{GD}. In the Meyer model, the mobile charge Q_I in the inversion layer is calculated in terms of V_{GS} and V_{GD}. C_{GS} and C_{GD} are found from $\dfrac{\partial Q_I}{\partial V_{GS}}$ and $\dfrac{\partial Q_I}{\partial V_D}$. We use Eq. (6.8) to find Q_I making the assumption that Q_{BO} is independent of y. This implies that V_T is a constant. We can integrate Eq. (6.8) to get Q_I as follows:

$$Q_I = \int_0^l (V_G - V_T - V(y)) C_{OX} W \, dy$$

$$= (V_G - V_T) C_{OX} W l - W C_{OX} \int_{V_S}^{V_D} \left(V(y) \frac{dy}{dV} \right) dV$$

where all voltages are with respect to the substrate. But

$$\frac{dy}{dV} = \frac{(V_G - V_T - V(y)) W C_{OX} \mu_n}{I_{DS}}$$

Substituting we get

$$Q_I = (V_G - V_T) C_{OX} W l - W C_{OX} l \int_{V_S}^{V_D} \frac{V(y)[V_G - V_T - V(y)] \, dV}{(V_G - V_T)(V_D - V_S) - \left(\dfrac{V_D^2}{2} - \dfrac{V_S^2}{2} \right)}$$

The expression for I_{DS} used above reduces to that in Eq. (6.9) for $V_S = 0$. Integrating and simplifying we get,

$$Q_I = \frac{2}{3} C_{OX} W l \left[\frac{(V_{GD} - V_T)^3}{(V_{GD} - V_T)^2 - (V_{GS} - V_T)^2} \right.$$

$$\left. - \frac{(V_{GS} - V_T)^3}{(V_{GD} - V_T)^2 - (V_{GS} - V_T)^2} \right]$$

$$\frac{\partial Q_I}{\partial V_{GS}} = C_{GS} = \frac{2}{3} C_{OX} W l \left[1 - \frac{(V_{GD} - V_T)^2}{(V_{GS} - V_T + V_{GD} - V_T)^2} \right]$$

$$(6.13)$$

$$\frac{\partial Q_I}{\partial V_{GD}} = C_{GD} = \frac{2}{3} C_{OX} W l \left[1 - \frac{(V_{GS} - V_T)^2}{(V_{GS} - V_T + V_{GD} - V_T)^2} \right]$$

$$(6.14)$$

Figure 6.21(a) shows the relevant capacitances when an inversion layer is present. C_{GS} and C_{GD} are plotted as a function of V_{DS} in Fig. 6.21(b). When $V_{DS} = 0$, $C_{GS} = C_{GD} = \frac{1}{2} C_{OX} W l$ as one would expect. For $V_{DS} = (V_{GS} - V_T)$, C_{GD} tends to zero and C_{GS} tends to $\frac{2}{3} C_{OX} W l$. The expression above is valid only in the region $V_{DS} \leqslant (V_{GS} - V_T)$. For $V_{DS} > (V_{GS} - V_T)$

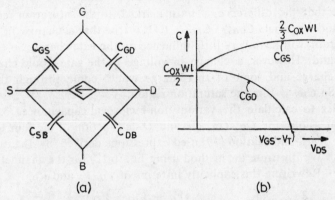

Fig. 6.21 (a) MOSFET capacitances with an inversion layer,
(b) Variation of C_{GS} and C_{GD} based on the Meyer
model [20]

the values of C_{GS} and C_{GD} at $V_{DS} = (V_{GS} - V_T)$ can be used. It is natural
that C_{GD} tends to zero in the saturation region. A depletion region
surrounds the drain and the drain is isolated from the channel. The overhang component of C_{GD} is still present.

Figure 6.22 shows C_{GB}, C_{GD} and C_{GS} as a function of V_{GS} in the three
regions viz., cut off, saturation and linear. This figure is for the Meyer Model.

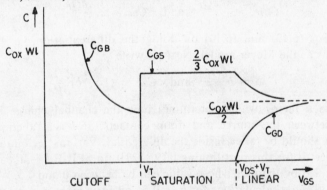

Fig. 6.22 Variation of various capacitances of a MOSFET as
a function of V_{GS}. This is based on the Meyer
model [20]

Ward and Dutton [19] have proposed a model based on non-reciprocal
capacitances. The non-reciprocal capacitances are (C_{GB}, C_{BG}), (C_{DG}, C_{GD})
and (C_{SG}, C_{GS}). Non-reciprocality means $C_{AB} \neq C_{BA}$ if A and B are the
terminals. New definitions have to be made for these capacitances.

$$C_{AB} = \frac{\partial Q_A}{\partial V_B} \quad \text{and} \quad C_{BA} = \frac{\partial Q_B}{\partial V_A}$$

C_{AB} describes the effect of a change in voltage at terminal B on the charge
at terminal A. Of the three pairs of non-reciprocal capacitances it is in C_{DG}

and C_{GD} that the difference is most marked. In the saturation region the Meyer model predicts $C_{GD} = C_{DG} = 0$. It is true that a change in voltage at the drain will have very little influence on the gate charge. So $C_{GD} = 0$ is still valid. However, a change in voltage at the gate would change the channel charge and some of this charge would come through the drain current. So $C_{DS} \neq 0$ in the saturation region.

In order to calculate the various non-reciprocal capacitances one must have expressions for Q_G, Q_B, Q_S and Q_D a function of the four terminal voltages. Ward and Dutton [19] used expressions derived by Ihantola and Moll [21]. We illustrate the method using Eq. (6.12) for the channel mobile charge Q_I. Rewriting this explicitly in terms of V_G, V_S and V_D,

$$Q_I = \frac{2}{3}C_{ox}Wl\left[\frac{(V_G - V_D - V_T)^3}{(V_G - V_D - V_T)^2 - (V_G - V_S - V_T)^2}\right.$$
$$\left.-\frac{(V_G - V_S - V_T)^3}{(V_G - V_D - V_T)^2 - (V_G - V_S - V_T)^2}\right]$$
$$= \frac{(V_G - V_D - V_T)^2 + (V_G - V_D - V_T)(V_G - V_S - V_T) + (V_G - V_S - V_T)^2}{[2(V_G - V_T) - V_D - V_S]}$$

$$(6.15)$$

If we assume that the part of gate charge Q_G depending on V_D and V_S is Q_I itself (but opposite in sign).

$$C_{GD} = -\frac{\partial Q_I}{\partial V_D} \quad \text{and} \quad C_{GS} = -\frac{\partial Q_I}{\partial V_S}$$

We get Eqs (6.13) and (6.14) on doing the differentiation, i.e. the same expressions as the Meyer model. Next we write

$$C_{DG} = b\frac{\partial Q_I}{\partial V_G} \quad \text{and} \quad C_{SG} = (1 - b)\frac{\partial Q_I}{\partial V_G}$$

where b is a factor which determines how the channel charge is to be divided between the source and drain contacts. If b is a function of the voltages it should be taken inside the differential. Various techniques have been suggested for this partitioning ([19], [22]). In SPICE b is a parameter XQC specified by the user whose value can be between 0 and 0.5. Ward and Dutton [19] show calculations for $b = 0.5$ and for a more accurate numerical method where b is a function of the terminal voltages. It has also, been suggested that when the gate voltage is increasing, the extra channel charge comes from the source and $b = 0$ [23]. Similarly when it is decreasing $b = 1$ on the drain supplies charge.

We can use Eq. (6.15) for Q_I and find C_{DG} and C_{SG} by putting $b = \frac{1}{2}$.

$$C_{DG} = b\frac{\partial Q_I}{\partial V_G}, \quad C_{SG} = (1 - b)\frac{\partial Q_I}{\partial V_G}$$

For $b = \frac{1}{2}$, we get

$$C_{DG} = C_{SG} = \frac{1}{3}C_{ox}Wl\left[1 + \frac{2(V_{GS} - V_T)(V_{GD} - V_T)}{(V_{GS} - V_T + V_{GD} - V_T)^2}\right]$$

At $\quad V_{GD} = V_T, C_{DG} = C_{SG} = \frac{1}{3}C_{ox}Wl$

In the saturation region $C_{DG} = \frac{1}{3} C_{OX} W$ and not 0 as predicted by the Meyer model while C_{SG} is half the value predicted by the Meyer model. C_{DG}, C_{SG}, C_{GD} and C_{GS} are plotted in Fig. 6.23.

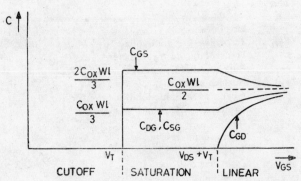

Fig. 6.23 Variation of C_{GS}, C_{SG}, C_{GD}, C_{DG} as a function of V_{GS} based on the non-reciprocal model of Ward and Dutton [19]

6.4.3 SPICE MOSFET Model

Table 6.4 gives the SPICE MOSFET parameters and typical values (default is NMOS). Note that some parameters can range over several orders of magnitude and typical values are more like examples.

The level of complexity is specified by the LEVEL parameter. LEVEL = 1 is the Schichman-Hodges model described by Eqs (6.9). Quantities like *VTO* and *KP* can be directly specified or calculated from *NSUB*, *TOX*, *W*, *L*, etc. The width *W* and the length *L* (*l* used here) are specified separately in the device card. LEVEL = 2 uses the more accurate equations given by Eq. (6.11). Also, mobility degradation is taken into account. LEVEL = 1 use the Meyer model to find C_{GD} and C_{GS} (but only if *TOX* is specified). For LEVELS = 2 and 3, one can use the Meyer model or the Ward-Dutton model. For *XQC* > 0.5, the Meyer model is used. For *XGC* ⩽ 0.5, *XQC* is used to partition the channel charge as described in the last section. The depletion capacitance C_{BD} and C_{BS} can be specified in one of two ways. One can directly give the zero bias value. Alternatively over can give the zero-bias values per unit area for the bottom of source/drain diffusion region (*CJ*) and the zero bias sidewall value per unit length of perimeter. This is shown in Fig. 6.24.

$$\text{Total bulk to drain capacitance} = \frac{AD \times CJ}{\left(1 + \frac{V_{DB}}{PB}\right)^{MJ}} + \frac{PD \times CJWS}{\left(1 + \frac{V_{DB}}{PB}\right)^{MJSW}}$$

AD, *PD*, *AS* and *PS* are specified in the **device card** along with *W* and *L*. Similarly the diode representing the reverse biased *D − B* and *S − B* junctions has a reverse saturation current *IS*. But it can also be specified per unit area as *JS*. Similarly one can specify *RD* and *RS* directly or the unit sheet resistance *RSH*.

Table 6.4 SPICE MOSFET Parameters

SPICE Parameter	Symbol used here	Description	Units	Default	Typical
LEVEL	—	Level of model	—	—	—
VTO	V_{TO}	Zero-bias threshold voltage	V	0	1.0
KP	β	Transconductance parameter	A/V²	2.0×10^{-5}	2.0×10^{-5}
GAMMA	γ	Bulk threshold parameter	V$^{1/2}$	0.0	0.3
PHI	$2\phi_F$	Surface potential	V	0.6	0.6
LAMBDA		Channel length modulation parameter (MOS1 and MOS2) only	1/V	0.0	0.02
RD	—	Drain ohmic resistance	Ohm	0.0	1.0
RS	—	Source ohmic resistance	Ohm	0.0	1.0
CBD	—	Zero-bias B-D jn. capacitance	F	0.0	20 FF
CBS	—	Zero-bias B-S jn. capacitance	F	0.0	20 FF
IS	—	Bulk jn. saturation current	A	1.0×10^{-14}	1.0×10^{-15}
CGSO	—	Gate-source overlap capacitance per meter channel width.	F/M	0.0	4.0×10^{-11}
CGDO	—	Gate-drain overlap capacitance per metre channel width	F/M	0.0	4.0×10^{-11}
CGBO	—	Gate-bulk overlap capacitance per meter channel length	F/M	0.0	2.0×10^{-10}
RSH	—	Drain and source diffusion sheet resistance	Ohm/sq	0.0	10.0
CJ	—	Zero-bias bulk jn. bottom capacitance per sq. meter of junction area	F/m²	0.0	2×10^{-4}
MJ	—	Bulk jn. bottom grading coefficient	—	0.4	0.5
CJSW	—	Zero-bias bulk jn. sidewall capacitance	F/M	0.0	1.0×10^{-9}
MJSW	—	Bulk junction sidewall grading coefficient	—	0.33	0.33
JS	—	Bulk junction saturation current per sq. meter of jn. area	A/m²	—	1.0×10^{-8}
TOX	t_{ox}	Oxide-thickness	meter	1.0×10^{-7}	1.0×10^{-75}
NSUB	N_A	Substrate doping	1/cm³	0.0	4.0×10^{-1}
NSS	Q_{ok}/q	Surface state density	1/cm²	0.0	1.0×10^{10}
NFS		Fast surface state density	1/cm²	0.0	1.0×10^{10}
TPG	—	Type of gate	—	—	—
		+1 opposite to substrate	—	—	—
		−1 same as substrate	—	—	—
		0 Al. gate	—	—	—
XJ	—	Metallurgical junction depth	m	0.0	1.0×10^{-6}
LD	—	Lateral diffusion	m	0.0	0.5×10^{-6}

SPICE Parameter	Symbol used here	Description	Units	Default	Typical
UO	$\mu_n(o)$	Surface mobility	cm²/V-S	600	700
UCRIT	E_{Crit}	Critical field for mobility degradation (MOS2 only)	V/cm	1.0×10^4	1.0×10^5
UEXP	U	Critical field exponent in mobility degradation (MGS2 only)	—	0.0	0.2
UTRA	—	Transverse field mobility coefficient (deleted for MOS2)	—	0.0	0.3
VMAX	S_{max}	Maximum drift velocity of carriers	m/s	0.0	1.0×10^5
NEFF	$NEFF$	Total channel charge coefficient (fixed and mobile) coefficient (MOS2 only)	m/s	0.0	1.0×10^5
XQC	b	Thin oxide model flag and coefficient of channel charge share attributed to drain (0 to 0.5) (XQC greater than 0.5 implies Meyer model)	—	1.0	0.4
KF	K_f	Flicker noise coefficient	—	0.0	1.0×10^{-26}
AF	K_f	Flicker noise exponent	—	1.0	1.2
FC	—	Coefficient for forward bias depletion capacitance formula	—	0.5	—
DELTA	—	Width effect on threshold voltage (MOS2 and MOS3 only)	—	0.0	1.0
THETA	θ	Mobility modulation MOS3 only)	1/V	0.0	0.1
ETA	—	Static feedback (MOS3 only)	—	0.0	1.0
KAPPA	—	Saturation field factor (MOS3 only)	—	0.2	0.5

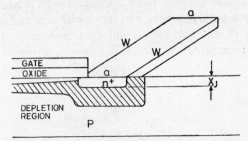

PERIMETER PD $= 2a + 2W$
AREA AD $= aW$

Fig. 6.24 Definition of PD and AD for calculation of sidewall capacitances in SPICE

The LEVEL = 3 model has a number of empirical parameters DELTA, THETA, ETC and KAPPA. THETA and KAPPA have been described in Section 6.4.1. DELTA and ETA account for second order phenomena described in Ref. [12]. X_j refers to the junction depth (see Fig. 6.24) and important for short channel effects [12]. One important effect not discussed here is sub-threshold conduction. When the channel potential (for $V_{SB} = 0$) is between ϕ_F and $2\phi_F$, a weak inversion layer exists. A small drain current proportional to a exp (ϕ_S/V_T) is then produced where ϕ_S is the channel potential. For details the reader is referred to Ref. [12] again.

A circuit representation of the SPICE MOSFET model is shown in Fig. 6.25. For convenience of representation, non-linear capacitances have been shown as linear.

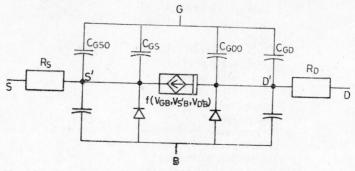

Fig. 6.25   Complete MOSFET model used in SPICE, C_{GSO} and C_{GDO} are overhang capacitances. All other capacitances are non-linear.

6.5 Noise Models for Semiconductor Devices

The kinds of noise usually accounted for are thermal noise, shot noise and flicker noise. Thermal noise and shot noise can be incorporated in a program like SPICE without the user having to specify any parameters, while flicker noise requires at least two parameters.

Any resistor R has thermal noise associated with it which can be represented either by a voltage source v in series with a noiseless resistor R or a current source i in parallel with a noiseless resistor R. The mean square values of v and i are given by

$$v^2 = 4kTR\,\Delta f$$

$$i^2 = \frac{4kT\,\Delta f}{R}$$

where

Δf = bandwidth of interest
k = Boltzmann constant
T = Temperature in °K.

Thermal noise is independent of frequency (and is therefore white) and bias currents. Thermal noise is easily incorporated in any device model by adding these current or voltage sources to any resistor. For example, a diode has a contact bulk resistance R_s (Fig. 6.1). The noise associated with it is a voltage source in series $v^2 = 4kTR_s \, \Delta f$. In a BJT, thermal noise is taken into account through the resistances R_B, R_C and R_E.

Shot noise is due to a certain randomness associated with the movement of holes and electrons across a depletion region. A steady DC current actually has minor fluctuations because of shot noise. Shot noise directly depends on bias currents. The mean square value of noise in the current is given by

$$i^2 = 2qI \, \Delta f$$

where I is the DC current and q the electronic charge. Though not indicated by the above equation shot noise decreases with frequency.

The origins of flicker noise are varied and rather complex. It is also dependent on bias current and independent of temperature. It is given by

$$i^2 = \frac{K_f I A_f \, \Delta f}{f^n}$$

where,

$\quad K_f =$ flicker noise coefficient

$\quad A_f =$ flicker noise exponent

$\quad n =$ constant often taken as unity (as in SPICE)

K_f and A_f correspond to parameters KF and AF in SPICE. Flicker noise varies as $1/f$ (for $n = 1$) and is also referred to as $1/f$ noise. The determination of KF is often difficult as it varies drastically from one device to another.

6.6 Summary

A circuit analysis program is of little use if it does not have built-in models for common semiconductor devices. Further, the user of these programs must be familar with the models built in a program before he can use the program effectively. In this chapter models for diodes, BJTs and MOSFETs present in SPICE have been described.

References

1. D.A. Calahan, Computer-aided Network Design, Chap. 5, McGraw-Hill Book Co., 1972.
2. C.I. Sah. R.N. Noyce and W. Shockley, Carrier Generation and Recombination in *p-n* Junction and *p-n* Junction Characteristics, Proc. IRE, 45, pp. 1228-1243, 1957.
3. S.M. Sze, Physics of Semiconductor Devices, p. 105, John Wiley and Sons, Inc., 1969.
4. D.A. Hodges and H.G. Jackson, Analysis and Design of Digital Integrated Circuits, Chap. 4, McGraw-Hill, Inc., 1983.
5. S.M. Sze, Physics of Semiconductor Devices, p. 102, John Wiley and Sons, Inc., 1969.

6. A. Vladimirescu, K. Zhang. A.R. Newton, D.O. Pederson, A.L. Sangiovanni-Vincentelli, SPICE Version 2G Users Guide, Dept, of Electrical Engineering and Computer Science, Univ. of California, Berkeley, 1981.

7. J.J. Ebers and J.L. Moll, Large-Signal Behaviour of Junction Transistors, Proc. IRE, 42, pp. 1761-1772, 1954.

8. S.M. Sze, Physics of Semiconductor Devices, p. 304, John Wiley and Sons, Inc., 1969.

9. H.K. Gummel and H.C. Poon, An Integral Charge Control Model of Bipolar Transistors, BSTJ 49, pp. 827-852, May, 1970.

10. P.R. Gray and R.G. Meyer, Analysis and Design of Analog Integrated Circuits, pp. 95-97, John Wiley and Sons, 1984.

11. H. Schichman and D.A. Hodges, Modelling and Simulation of IGFET Switching Circuits, IEEE J. SSC, SC-C, pp. 285-289, Sept. 1968.

12. L.A. Glasser and D.W. Dobberpuhl, The Design and Analysis of VLSI Circuits, Chap. 2, Addison-Wesley Publishing Co., 1985.

13. P. Richman, Characteristics and Operation of MOS Field-Effect Devices, Chaps. 3 and 4, McGraw-Hill Book Co., 1967.

14. S.M. Sze, Physics of Semiconductor Devices, p. 444, John Wiley and Sons, Inc., 1969.

15. G. Baum and H. Beneking, Drift Velocity Saturation in MOS Transistors, IEEE Trans. Electron Devices, ED-17, pp. 481-482, 1970.

16. A.G. Sabnis and J.T. Clements, Characterization of Electron Mobility in the Inverted (100) Si Surface IEEE International Electron Device Meeting, pp. 18-21, Washington DC, 1979.

17. P. Smith, M. Inoue and J. Frey, Electron Velocity in Si and GaAs at Very High Electric Fields, Appl. Phys. Lett, 37, pp. 797-798, 1968.

18. C. Jacobini, C. Canali, G. Ottaviani and A.A. Quaranta, A Review of some Charge Transport Properties in Silicon, Solid State Electronics, 20, pp. 77-89, 1977.

19. D.F. Ward and R.W. Dutton, A Charge-Oriented Model for MOS Transistor Capacitances, IEEE, J. Solid State Ccts, SC-13, pp. 703-707, 1978.

20. J.E. Meyer, MOS Models and Circuits Simulation, RCA Review, 32, pp. 42-63. March, 1971.

21. H.K.J. Ihantola and J.M. Moll, Design theory of a surface field effect transistor, Solid State Electronics, 7, pp. 423-430, June, 1964.

22. G.M. Taylor, W. Fichtner and J.G. Simmons, A Description of MOS Internodal Capacitances for Transient Simulations, IEEE Trans. CAD, pp. 150-156, Oct. 1982.

23. J.J. Paulos and D.A. Antoniadis, Limitations of Quasi-Static Models for the MOS Transistor, IEEE Trans. Electron Devices Lett., EDL-4, pp. 221-224, 1983.

Problems

1. In pn junction diode $N_A = 2*10^{20}$ cm^{-3} and $N_D = 5*10^{17}$ cm^{-3}. In the p region, $D_n = $ cm^2/s and $L_n = 10$ μ and in the n region, $D_p = 4$ cm^2/s and $L_p = 5$ μ. The area is 100 sq. microns and the grading coefficient is 0.5. Find
   (a) The barrier potential
   (b) The depletion region charge, the depletion region width and maximum electric field at $V = -5$ V and $+0.5$ V.
   (c) The saturation current
   (d) The depletion capacitance and diffusion capacitance at $V = -5$V and 0.7V.

2. Repeat Problem 1 where the diode is short based with the width of the n region as 1 μm.

3. The following are the Ebers-Moll parameters of a transistor. Calculate I_B and I_E. Also calculate the various capacitances. Assume $V_{BE} = 0.8$ v, $V_{BC} = 0.65$ V,

$I_{EFS} = 10^{-12}$ A, $I_{CFS} = 2*10^{-15}$ A, $\alpha_N = 0.99$. Both emission coefficients $= 1$. Both grading coefficients $= 0.5$, $\tau_{tf} = 0.1$ n secs, $\tau_{tr} = 10$ n secs. Zero bias depletion capacitances of both junctions $= 20$ pf. Use default values for other parameters.

4. In a symmetrical Si *npn* transistor, the following currents were measured at the two sets of voltages given

   (a) $V_{BE} = 0.7$ V, $V_{BC} = -2$ V; $I_C = 2$ mA, $I_B = 200$ μA.

   (b) $V_{BE} = 0.1$ V, $V_{BC} = -2$ V; $I_C = 3.6*10^{-9}$ mA.

   Find the parameters of the DC transport model to represent this tansistor. Draw a figure.

5. Calculate the parameters of the transport model corresponding to the Ebers-Moll parameters of Problem 3. Assume the reverse saturation current of the non-ideal diodes to be 100 I_S.

6. Show that the transport model, linearised at some operating point in the normal active mode, produces the hybrid-π model. Calculate the parameters of the hybrid-π model for the operating point specified below. Use the transport model of problem 5. Assume $V_{BE} = 0.75$ V and $V_{BC} = -4$ V for the operating point. Point out one feature of the hybrid-π model not represented by the transport model.

7. The following parameters are specified for the DC Gummel-Poon model:

   $I_s = 10^{-12}$ mA, $I_{SC} = I_{SE} = 10^{-11}$ mA, $V_{AF} = V_{AR} = 100$ V,

   $I_{KF} = I_{KR} = 50$ mA, $\eta_F = \eta_R = 1$, $\eta_C = \eta_E = 2$, $\beta_F = 100$, $\beta_R = 0.1$

   Calculate I_C and I_B when (a) $V_{BE} = 0.75$, $V_{CE} = 0.15$ V and

   (b) $V_{BE} = 0.01$ V, $V_{BC} = -10$ V.

8. A NPN transistor is operating in the normal active mode. The ratio I_C/I_B is 100 at $I_C = 5$ mA and $V_{BE} = 0.7$ V and increases to 150 at $I_C = 10$ mA. Find IS ISE and BF corresponding to the SPICE Gummel-Poon model. State your assumptions.

9. Consider a transistor in the normal active mode. If I_C is far less than I_{KF},

$$Q_2 = \frac{I_s}{I_{KF}} \exp (V_{BE}/VT)$$

Neglecting Early effect,

$$Q_B = \frac{1}{2} [1 + (1 + 4Q_2)^{1/2}] \simeq \frac{1}{2}[1 + 1 + 2Q_2] \simeq [1 + Q_2)$$

or $$\frac{1}{Q_B} \simeq (1 - Q_2) = \left(1 - \frac{I_s}{I_{KF}} X^2 \right) \text{ where } x = \exp (V_{BE}/2V_T)$$

Show that the ratio I_C/I_B has a maximum at the value of x given by

$$I_{SE} = \frac{2I_s^2 x^3}{\beta_F I_{KF}} + \frac{3I_{SE}I_s x^2}{I_{KF}}$$

10. Assume that the mobile charge per unit area $Q_I(y)$ is given at some distance y from the source in the channel of a NMOS device by

$$Q_I(y) = a_2 V(y) + a_1 [V(y)]^{1/2} + a_0$$

   where $V(y)$ is the voltage at distance y and a_0, a_1 are constants corresponding to some gate voltage. Find the drain to source current I_{DS} as a function of V_D and V_S.

11. Find V_T for $V_{SB} = OV$ and $V_{SB} = 4$ V for a NMOS transistor with the following values: $NA = 5*10^{-15}$ cm^{-3}, $\phi_{GC} = -0.8$ V O, $2\phi_F = 0.6$ V, $Q_{OX} = q*5*10^{10}$ coulombs/cm^2, $\epsilon_{sl} = 11.7$, $\epsilon_{OX} = 3.9$

12. In deriving an expression for V_{TO}, the term Q_{BO} occurred. Let the expression for Q_{BO} be expanded using the Taylor series as

$$Q_{BO} = Q_{BO}(V_s) + (V(y) - V(s)) \frac{dQ_{BO}}{dV}\Big/_{V(y)\ =\ V_S}$$

Show that I_{DS} can then be written as

$$I_{DS} = \beta[(V_{GS} - V_T) V_{DS} - (1 + \delta) V_{DS}^2/2]$$

for $V_{DS} > V_D$ sat where

$$V_D \text{ sat} = \frac{(V_{GS} - VT)}{(1 + \delta)} \quad \text{and} \quad \delta = r/2 \ (V_{SF} + 2\phi_F)^{1/2}$$

13. The following are the SPICE Parameters for a NMOS transistor.
$KP = 20 \ \mu A/V^2$, $VTO = 1$ V, GAMMA $= 0.4 \ V^{1/2}$
$PHI = 0.6 \ V$, LAMBDA $= 0.025$, $W/L = 10$ (W and L have to be given individually on the device 'card'). Find I_{DS} for

(a) $V_S = V_B = 0$, $V_{DS} = 0.5$ V, $V_{GS} = 5$ V and

(b) $V_B = 0$, $V_{SB} = 3$ V, $V_{DS} = 5$ V, $V_{GS} = 3$ V.

7

Implementation and Use of a General Purpose Circuit Analysis Program Like Spice

The first five chapters described the principles behind circuit analysis. Algorithms for both formulation and solution of the relevant equations were studied. Chapter 6 described models for commonly used semiconductor devices. Here we discuss how all these ideas can be put together in a computer program. Applications using the popular program SPICE ([1], [2]) are given.

7.1 Constituent Parts of a Circuit Analysis Program

Any circuit analysis program would have the following important sections:

1. Parser for input file
2. Equation formulation routines
3. Routines to solve a system of sparse linear equations
4. Models for semiconductor devices

The kind of input description given in Chapter 2 is very restrictive. The description should preferably be free format for user convenience. Therefore each line of the input file is read as a character string ('A' format in FORTRAN). Each line is analysed or parsed as a compiler would and its information extracted. The input data specification then takes on the nature of a computer language. While this makes the input format quite flexible, the parser may take up a considerable part of the total code.

A program with sub-circuit facility will help a user considerably. If a circuit segment occurs in many places, it may be described just once in the input file and called at different times with different node numbers. This sub-circuit facility makes the input file very compact. One can even have sub-circuits within sub-circuits so that the description builds up hierarchically. Where sub-circuits exist they must first be expanded before the analysis can proceed. This may be somewhat like expansion of macros by a macro-assembler. That is extra nodes have to be created and the nodes now renumbered. Also, built-in models for elements like BJTs have to be incorporated in the same way as user defined sub-circuit elements.

It may be stressed again that a routine for solving sparse linear equations is the heart of the program. This should be written as efficiently as possible.

7.2 Examples Using Spice

The use of the program SPICE is described using some examples.

The first example is the TTL inverter input-output characteristic. The circuit is shown in Fig. 7.1. Proper parameters have to be chosen for each of the transistors. The diode is a transistor with the $C–B$ junction shorted. The analysis is DC and therefore one need not worry about capacitances.

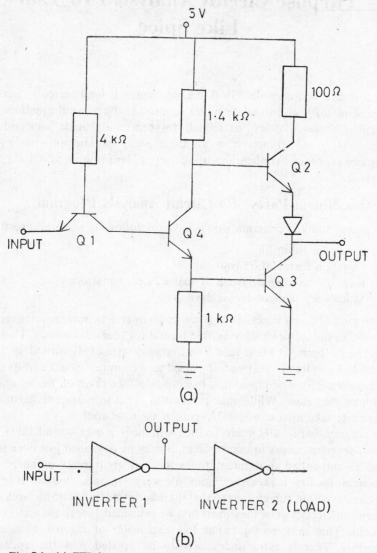

Fig. 7.1   (a) TTL inverter (b) Input-output characteristic of inverter 1 is calculated using inverter 2 as load.

Parameters chosen here are $\beta_F = 70$, $\beta_R = 1$, $I_S = 2.1 \times 10^{-16}$ A. Computer printouts of the input file and the plotted output are shown in Fig. 7.2(a) and Fig. 7.2(b) respectively. The inverter has a fanout of 1 and the inverter

```
1*******17 Dec 8 ********  SPICE 2G.6    3/15/83 *********11:43:49*****

0*TTL INVERTER INPUT-OUTPUT CHARACTERISTIC*

0****     INPUT LISTING                  TEMPERATURE =   27.000 DEG C

0***************************************************************************


    VCC 2 0 5V
    VIN 1 0
    .SUBCKT INV 1 2 3
    Q1 5 4 1 M1
    Q2 8 6 9 M1
    Q3 3 7 0 M1
    Q4 6 5 7 M1
    Q5 9 9 3 M1
    R1 2 4 4K
    R2 2 6 1.4K
    R3 7 0 1K
    R4 2 8 100
    .MODEL M1 NPN
    .ENDS INV
    X1 1 2 4 INV
    X2 4 2 3 INV
    .DC VIN 0 5 0.2
    .PLOT DC V(4), V(1)
    .END
```

(a)

```
1****************17 Dec 8 ********************  SPICE 2G.6    3/15/83 ***************************11:43:49****************

0*TTL INVERTER INPUT-OUTPUT CHARACTERISTIC*
0****                   DC TRANSFER CURVES                          TEMPERATURE =   27.000 DEG C

0**********************************************************************************************************************

0LEGEND:

*: V(4)
+: V(1)
X
    VIN        V(4)

(*)--------------    0.000D+00       1.000D+00       2.000D+00       3.000D+00       4.000D+00

(+)--------------    0.000D+00       2.000D+00       4.000D+00       6.000D+00       8.000D+00

0.000D+00   3.465D+00  +                       .               .               .               *     .
2.000D-01   3.465D+00  . +                     .               .               .               *     .
4.000D-01   3.465D+00  .   +                   .               .               .               *     .
6.000D-01   3.462D+00  .     +                 .               .               .             *       .
8.000D-01   3.323D+00  .       +               .               .               .           *         .
1.000D+00   3.081D+00  .         +             .               .               .         *           .
1.200D+00   2.823D+00  .           +           .               .             *                       .
1.400D+00   2.557D+00  .             +         .               .           *                         .
1.600D+00   1.721D+00  .               +       .             *                                       .
1.800D+00   2.169D-02  .*                +     .                                                     .
2.000D+00   2.168D-02  .*                  +   .                                                     .
2.200D+00   2.169D-02  .*                    + .                                                     .
2.400D+00   2.169D-02  .*                      +                                                     .
2.600D+00   2.169D-02  .*                      . +                                                   .
2.800D+00   2.169D-02  .*                      .   +                                                 .
3.000D+00   2.169D-02  .*                      .     +                                               .
3.200D+00   2.169D-02  .*                      .       +                                             .
3.400D+00   2.169D-02  .*                      .         +                                           .
3.600D+00   2.169D-02  .*                      .           +                                         .
3.800D+00   2.169D-02  .*                      .             +                                       .
4.000D+00   2.169D-02  .*                      .               +                                     .
4.200D+00   2.169D-02  .*                      .                 +                                   .
4.400D+00   2.169D-02  .*                      .                   +                                 .
4.600D+00   2.169D-02  .*                      .                     +                               .
4.800D+00   2.169D-02  .*                      .                       +                             .
5.000D+00   2.169D-02  .*                      .                         +                           .

Y
0
0      JOB CONCLUDED
       TOTAL JOB TIME       2.93
```

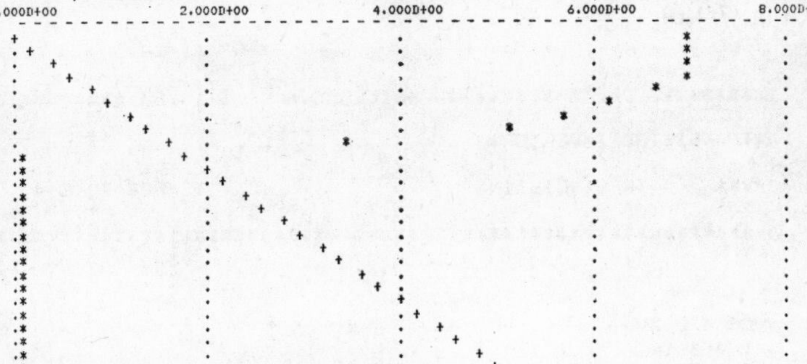

(b)

Fig. 7.2   SPICE analysis of TTL inverter of Fig. 7.1. (a) SPICE input file (b) Plot of I/O characteristic.

is defined as sub-circuit. The sub-circuit is called twice, once as a load and once as an element whose input-output characteristic is to be calculated.

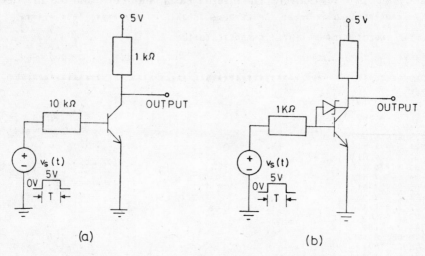

Fig. 7.3 Simple BJT inverter (b) BJT inverter with Schottky diode.

As a second example, consider the transistor inverter shown in Fig. 7.3(a). A pulse input is given to it as shown at the base. Figure 7.4(a) shows the input file and Fig. 7 4(b) shows the SPICE output plot when transient analysis is done. The Schottky transistor introduced in Fig. 7.3(b) prevents the BJT from going into saturation. Correspondingly, the SPICE output in Fig. 7.4(d) shows a drastic reduction in storage time. Representation of capacitances in both the BJT and Schottky diode is very important for transient analysis. Note how the parameters are given to describe the various diffusion and depletion capacitances in the input files in Fig. 7.4(a) and Fig. 7.4(c).

```
1*******17 Dec 8 ********  SPICE 2G.6   3/15/83 ********16:45:19*****

0*TRANSISTOR INVERTER*

0****     INPUT LISTING               TEMPERATURE =   27.000 DEG C

0************************************************************************


    VCC 4 0 5V
    R1 4 3 1K
    R2 1 2 10K
    Q1 3 2 0 M1
    .MODEL M1 NPN BF=70 BR=1 IS=2E-16 TF=0.2NS TR=20NS CJE=0.3PF
    +VJE=0.9V MJE=0.5 CJC= 0.15PF VJC=0.7V MJC=0.33
    VIN 1 0 PULSE(0V 5V 2NS 2NS 2NS 20NS 70NS)
    .TRAN 1NS 67NS
    .PLOT TRAN V(3), V(1)
    .END
```

(a)

```
1*****************17 Dec 8 ********************** SPICE 2G.6   3/15/83 ***************************16:45:19***

0*TRANSISTOR INVERTER*                                          TEMPERATURE =   27.000 DEG C
0****          TRANSIENT ANALYSIS

0**********************************************************************************************************

0LEGEND:

*: V(3)
+: V(1)
X
      TIME       V(3)
    (**)--------------  0.000D+00          2.000D+00          4.000D+00          6.000D+00
                                                                              |
   0.000D+00   5.000D+00   +                                             *    |
   1.000D-09   5.000D+00   +                                             *    |
   2.000D-09   5.000D+00   +                                              *   |
   3.000D-09   5.035D+00   .                                         *    +   |
   4.000D-09   4.746D+00   .                                              +   |
   5.000D-09   3.438D+00   .                        *                     +   |
   6.000D-09   2.186D+00   .              *                               +   |
   7.000D-09   1.019D+00   .        *                                     +   |
   8.000D-09   2.169D-01   . *                                            +   |
   9.000D-09   1.537D-01   . *                                            +   |
   1.000D-08   1.328D-01   . *                                            +   |
   1.100D-08   1.226D-01   . *                                            +   |
   1.200D-08   1.159D-01   . *                                            +   |
   1.300D-08   1.106D-01   .*                                             +   |
   1.400D-08   1.064D-01   .*                                             +   |
   1.500D-08   1.030D-01   . *                                            +   |
   1.600D-08   1.003D-01   .*                                             +   |
   1.700D-08   9.784D-02   .*                                             +   |
   1.800D-08   9.571D-02   .*                                             +   |
   1.900D-08   9.386D-02   .*                                             +   |
   2.000D-08   9.225D-02   .*                                             +   |
   2.100D-08   9.079D-02   .*                                             +   |
   2.200D-08   8.947D-02   .*                                             +   |
   2.300D-08   8.828D-02   .*                                             +   |
   2.400D-08   8.720D-02   .*                                             +   |
   2.500D-08   8.588D-02   .*                                             +   |
   2.600D-08   8.568D-02   +*                                                 |
   2.700D-08   8.766D-02   +*                                                 |
   2.800D-08   9.007D-02   +*                                                 |
   2.900D-08   9.257D-02   +*                                                 |
   3.000D-08   9.525D-02   +*                                                 |
   3.100D-08   9.810D-02   +*                                                 |
   3.200D-08   1.011D-01   +*                                                 |
   3.300D-08   1.044D-01   +*                                                 |
   3.400D-08   1.080D-01   +*                                                 |
   3.500D-08   1.120D-01   +*                                                 |
   3.600D-08   1.164D-01   +*                                                 |
   3.700D-08   1.213D-01   + *                                               5|
   3.800D-08   1.275D-01   + *                                                |
   3.900D-08   1.351D-01   + *                                               2|
   4.000D-08   1.451D-01   + *                                                |
   4.100D-08   1.586D-01   + *                                                |
   4.200D-08   1.994D-01   + *                                                |
   4.300D-08   4.462D-01   +     *                                            |
   4.400D-08   8.097D-01   +        *                                         |
   4.500D-08   1.188D+00   +            +                                     |
   4.600D-08   1.602D+00   +              *                                   |
   4.700D-08   1.991D+00   +                *                                 |
   4.800D-08   2.367D+00   +                   *                              |
   4.900D-08   2.748D+00   +                     *                            |
   5.000D-08   3.092D+00   +                       *                          |
   5.100D-08   3.430D+00   +                         *                        |
   5.200D-08   3.757D+00   +                           *     *                |
   5.300D-08   4.055D+00   +                                 *                |
   5.400D-08   4.340D+00   +                                     *            |
   5.500D-08   4.594D+00   +                                       *          |
   5.600D-08   4.808D+00   +                                         *        |
   5.700D-08   4.961D+00   +                                          *       |
   5.800D-08   4.986D+00   +                                          *       |
   5.900D-08   4.993D+00   +                                          *       |
   6.000D-08   4.994D+00   +                                          *       |
   6.100D-08   4.994D+00   +                                          *       |
   6.200D-08   4.996D+00   +                                          *       |
   6.300D-08   4.997D+00   +                                          *       |
   6.400D-08   4.997D+00   +                                          *       |
   6.500D-08   4.998D+00   +                                          *       |
   6.600D-08   4.998D+00   +                                          *       |
   6.700D-08   4.999D+00   +- - - - - - - - - - - - - - - - - - - - - *- - - -|

Y
0
0           JOB CONCLUDED
            TOTAL JOB TIME        2.18
```

(b)

```
1********17 Dec 8 ********  SPICE 2G.6    3/15/83 ********16:51:22*****

0*TRANSISTOR INVERTER*

0****      INPUT LISTING                  TEMPERATURE =   27.000 DEG C

0**************************************************************************


    VCC 4 0 5V
    R1 4 3 1K
    R2 1 2 10K
    Q1 3 2 0 M1
    D1 2 3 M2
    .MODEL M2 D IS=0.7E-11 CJO=0.05PF VJ=0.7V M=0.5
    .MODEL M1 NPN BF=70 BR=1 IS=2E-16 TF=0.2NS TR=20NS CJE=0.3PF
    +VJE=0.9V MJE=0.5 CJC= 0.15PF VJC=0.7V MJC=0.33
    VIN 1 0 PULSE(0V 5V 2NS 2NS 2NS 20NS 70NS)
    .TRAN 1NS 67NS
    .PLOT TRAN V(3), V(1)
    .END
```

(c)

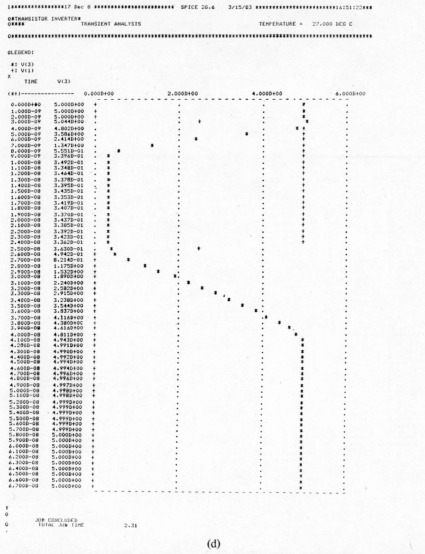

(d)

Fig. 7.4 (a) SPICE input file for BJT inverter (b) SPICE pulse response plot for BJT inverter (c) SPICE input file for Schottky transistor inverter. (d) SPICE pulse response plot for Schottky transistor inverter. Note the absence of storage time as compared to (b).

A NMOS inverter with a depletion mode transistor as load is shown in Fig. 7.5. The SPICE output is shown in Fig. 7.6. Parameters for the transistors are from Ref. [3] for a typical 2 μm process. Figure 7.6(b) shows the input-output characteristic and Fig. 7.6(d) the response to a pulse input. Figure 7.7(b) and (d) are for a similar inverter with the difference that the length of the depletion transistor is reduced to half its previous value. Notice the deterioration in the input-output characteristic and the improve-

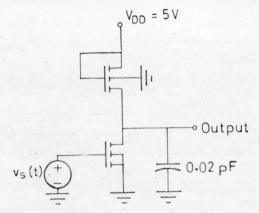

Fig. 7.5   NMOS inverter with depletion mode load.

ment in the pulse response. The input files for DC *I/O* characteristic of the NMOS inverters are in Figs 7.6(a) and 7.7(a). The input files for pulse response are in Figs 7.6(c) and 7.7(c).

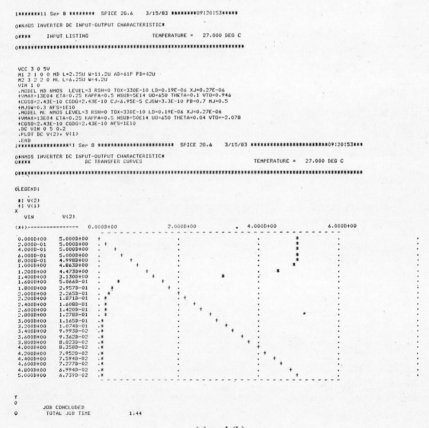

(a) and (b)

```
1*******17 Dec 8 *******   SPICE 2G.6    3/15/83 ********17:05:06*****

0*NMOS INVERTER PULSE RESPONSE*

0****     INPUT LISTING                    TEMPERATURE =   27.000 DEG C

0*****************************************************************************


VCC 3 0 5V
M1 2 1 0 0 MD L=2.25U W=11.2U AD=61P PD=42U
M2 3 2 2 0 ML L=3.125U W=4.2U
VIN 1 0 PULSE(0V 5V 2NS 2NS 2NS 10NS 30NS)
C1 2 0 0.02PF .
.MODEL MD NMOS  LEVEL=3 RSH=0 TOX=330E-10 LD=0.19E-06 XJ=0.27E-06
+VMAX=13E04 ETA=0.25 KAPPA=0.5 NSUB=5E14 UO=650 THETA=0.1 VTO=0.946
+CGSO=2.43E-10 CGDO=2.43E-10 CJ=6.95E-5 CJSW=3.3E-10 PB=0.7 MJ=0.5
+MJSW=0.3 NFS=1E10
.MODEL ML NMOS LEVEL=3 RSH=0 TOX=330E-10 LD=0.19E-06 XJ=0.27E-06
+VMAX=13E04 ETA=0.25 KAPPA=0.5 NSUB=50E14 UO=650 THETA=0.04 VTO=-2.078
+CGSO=2.43E-10 CGDO=2.43E-10 NFS=1E10
.TRAN 0.5NS 25NS
.PLOT TRAN V(2), V(1)
.END
```

(c)

```
1***************17 Dec 8 ***********************   SPICE 2G.6    3/15/83 ********************17:05:06***

0*NMOS INVERTER PULSE RESPONSE*
0****                TRANSIENT ANALYSIS                        TEMPERATURE =   27.000 DEG C

0*************************************************************************************************************

OLEGEND:

*: V(2)
+: V(1)
X
   TIME      V(2)

(*+)---------------   0.000D+00         2.000D+00         4.000D+00         6.000D+00
                    - - - - - - - - - - - - - - - - - - - - - - - - - - - - - - - - -
 0.000D+00   5.000D+00   +         .              .              *         .
 5.000D-10   5.000D+00   +         .              .              *         .
 1.000D-09   5.000D+00   +         .              .              *         .
 1.500D-09   5.000D+00   +         .              .              *         .
 2.000D-09   5.000D+00   +         .              .              *         .
 2.500D-09   4.941D+00   .        +               .              *         .
 3.000D-09   2.999D+00   .         .              +       *      .         .
 3.500D-09   2.785D-01   . *       .              .       +      .         .
 4.000D-09   1.760D-01   . *       .              .              +         .
 4.500D-09   1.414D-01   . *       .              .              +         .
 5.000D-09   1.412D-01   . *       .              .              +         .
 5.500D-09   1.412D-01   . *       .              .              +         .
 6.000D-09   1.412D-01   . *       .              .              +         .
 6.500D-09   1.412D-01   . *       .              .              +         .
 7.000D-09   1.412D-01   . *       .              .              +         .
 7.500D-09   1.412D-01   . *       .              .              +         .
 8.000D-09   1.412D-01   . *       .              .              +         .
 8.500D-09   1.412D-01   . *       .              .              +         .
 9.000D-09   1.412D-01   . *       .              .              +         .
 9.500D-09   1.412D-01   . *       .              .              +         .
 1.000D-08   1.412D-01   . *       .              .              +         .
 1.050D-08   1.412D-01   . *       .              .              +         .
 1.100D-08   1.412D-01   . *       .              .              +         .
 1.150D-08   1.412D-01   . *       .              .              +         .
 1.200D-08   1.412D-01   . *       .              .              +         .
 1.250D-08   1.412D-01   . *       .              .              +         .
 1.300D-08   1.412D-01   . *       .              .              +         .
 1.350D-08   1.412D-01   . *       .              .              +         .
 1.400D-08   1.412D-01   . *       .              .              +         .
 1.450D-08   1.453D-01   . *       .              .        +     .         .
 1.500D-08   2.383D-01   . *       .              +              .         .
 1.550D-08   9.693D-01   .         * +            .              .         .
 1.600D-08   2.317D+00   +         .        *     .              .         .
 1.650D-08   3.902D+00   +         .              .       *.     .         .
 1.700D-08   4.751D+00   +         .              .              *         .
 1.750D-08   4.987D+00   +         .              .              .*        .
 1.800D-08   5.002D+00   +         .              .              *         .
 1.850D-08   5.001D+00   +         .              .              *         .
 1.900D-08   5.000D+00   +         .              .              *         .
 1.950D-08   5.000D+00   +         .              .              *         .
 2.000D-08   5.000D+00   +         .              .              *         .
 2.050D-08   5.000D+00   +         .              .              *         .
 2.100D-08   5.000D+00   +         .              .              *         .
 2.150D-08   5.000D+00   +         .              .              *         .
 2.200D-08   5.000D+00   +         .              .              *         .
 2.250D-08   5.000D+00   +         .              .              *         .
 2.300D-08   5.000D+00   +         .              .              *         .
 2.350D-08   5.000D+0C   +         .              .              *         .
 2.400D-08   5.000D+0C   +         .              .              *         .
 2.450D-08   5.000D+0C   +         .              .              *         .
 2.500D-08   5.000D+00   +         .              .              *         .
                    - - - - - - - - - - - - - - - - - - - - - - - - - - - - - - - - -
Y
0
     JOB CONCLUDED
     TOTAL JOB TIME        3.45
0
.
```

(d)

Fig. 7.6 (a) Input file for NMOS inverter I/O characteristic. (b) SPICE I/O characteristic for NMOS inverter. (c) Input file for NMOS inverter pulse response. (d) SPICE pulse response for NMOS inverter.

```
1********11 Sep 8 ******** SPICE 2G.6    3/15/83 ********09:25:21*****

0*NMOS INVERTER DC INPUT-OUTPUT CHARACTERISTIC*

0****     INPUT LISTING                 TEMPERATURE =   27.000 DEG C

0***************************************************************************


VCC 3 0 5V
M1 2 1 0 0 MD L=2.25U W=11.2U AD=61P PD=42U
M2 3 2 2 0 ML L=3.125U W=4.2U
VIN 1 0
.MODEL MD NMOS  LEVEL=3 RSH=0 TOX=330E-10 LD=0.19E-06 XJ=0.27E-06
+VMAX=13E04 ETA=0.25 KAPPA=0.5 NSUB=5E14 UO=650 THETA=0.1 VTO=0.946
+CGSO=2.43E-10 CGDO=2.43E-10 CJ=6.95E-5 CJSW=3.3E-10 PB=0.7 MJ=0.5
+MJSW=0.3 NFS=1E10
.MODEL ML NMOS LEVEL=3 RSH=0 TOX=330E-10 LD=0.19E-06 XJ=0.27E-06
+VMAX=13E04 ETA=0.25 KAPPA=0.5 NSUB=50E14 UO=650 THETA=0.04 VTO=-2.078
+CGSO=2.43E-10 CGDO=2.43E-10 NFS=1E10
.DC VIN 0 5 0.2
.PLOT DC V(2), V(1)
.END
1**************11 Sep 8 *******************  SPICE 2G.6  .3/15/83 *****************************09:25:21***

0*NMOS INVERTER DC INPUT-OUTPUT CHARACTERISTIC*
0****                 DC TRANSFER CURVES                          TEMPERATURE =   27.000 DEG C

0****************************************************************************************************


0LEGEND:

*: V(2)
+: V(1)
X
   VIN        V(2)

(*+)---------------    0.000D+00        2.000D+00        4.000D+00        6.000D+00
0.000D+00    5.000D+00   +           .                .                *         .
2.000D-01    5.000D+00   . +         .                .                *         .
4.000D-01    5.000D+00   .   +       .                .                *         .
6.000D-01    5.000D+00   .     +     .                .                *         .
8.000D-01    4.999D+00   .       +   .                .                *         .
1.000D+00    4.939D+00   .         + .                .              *           .
1.200D+00    4.779D+00   .           +                .            *             .
1.400D+00    4.485D+00   .           . +              .          *               .
1.600D+00    3.882D+00   .           .   +            .      *.                   .
1.800D+00    1.802D+0(   .           .     +*         .                          .
2.000D+00    5.976D-01   .       *   .       +        .                          .
2.200D+00    4.433D-0:   .     *     .         +      .                          .
2.400D+00    3.653D-01   .   *       .           +    .                          .
2.600D+00    3.150D-01   .  *        .             +  .                          .
2.800D+00    2.791D-01   . *         c                .                          .
3.000D+00    2.520D-01   . *         .                +                          .
3.200D+00    2.307D-01   . *         .                . +                        .
3.400D+00    2.134D-01   . *         .                .   +                      .
3.600D+00    1.991D-01   . *         .                .     +                    .
3.800D+00    1.870D-0:   . *         .                .       +                  .
4.000D+00    1.766D-01   . *         .                .         +                .
4.200D+00    1.677D-01   . *         .                .                          .
4.400D+00    1.598D-01   . *         .                .                          .
4.600D+00    1.529D-01   . *         .                .             +            .
4.800D+00    1.467D-01   . *         .                .               +          .
5.000D+00    1.412D-01   . *         .                .                 +        .
```

Y
0
0   JOB CONCLUDED
            TOTAL JOB TIME   1.24

(a) and (b)

```
1********17 Dec 8 ******** SPICE 2G.6    3/15/83 ********16:58:45*****

0*NMOS INVERTER PULSE RESPONSE*

0****     INPUT LISTING                 TEMPERATURE =   27.000 DEG C

0****************************************************************************


VCC 3 0 5V
M1 2 1 0 0 MD L=2.25U W=11.2U AD=61P PD=42U
M2 3 2 2 0 ML L=6.25U W=4.2U
VIN 1 0 PULSE(0V 5V 2NS 2NS 2NS 10NS 30NS)
C1 2 0 0.02PF
.MODEL MD NMOS  LEVEL=3 RSH=0 TOX=330E-10 LD=0.19E-06 XJ=0.27E-06
+VMAX=13E04 ETA=0.25 KAPPA=0.5 NSUB=5E14 UO=650 THETA=0.1 VTO=0.946
+CGSO=2.43E-10 CGDO=2.43E-10 CJ=6.95E-5 CJSW=3.3E-10 PB=0.7 MJ=0.5
+MJSW=0.3 NFS=1E10
.MODEL ML NMOS LEVEL=3 RSH=0 TOX=330E-10 LD=0.19E-06 XJ=0.27E-06
+VMAX=13E04 ETA=0.25 KAPPA=0.5 NSUB=50E14 UO=650 THETA=0.04 VTO=-2.078
+CGSO=2.43E-10 CGDO=2.43E-10 NFS=1E10
.TRAN 0.5NS 25NS
.PLOT TRAN V(2), V(1)
.END
```

(c)

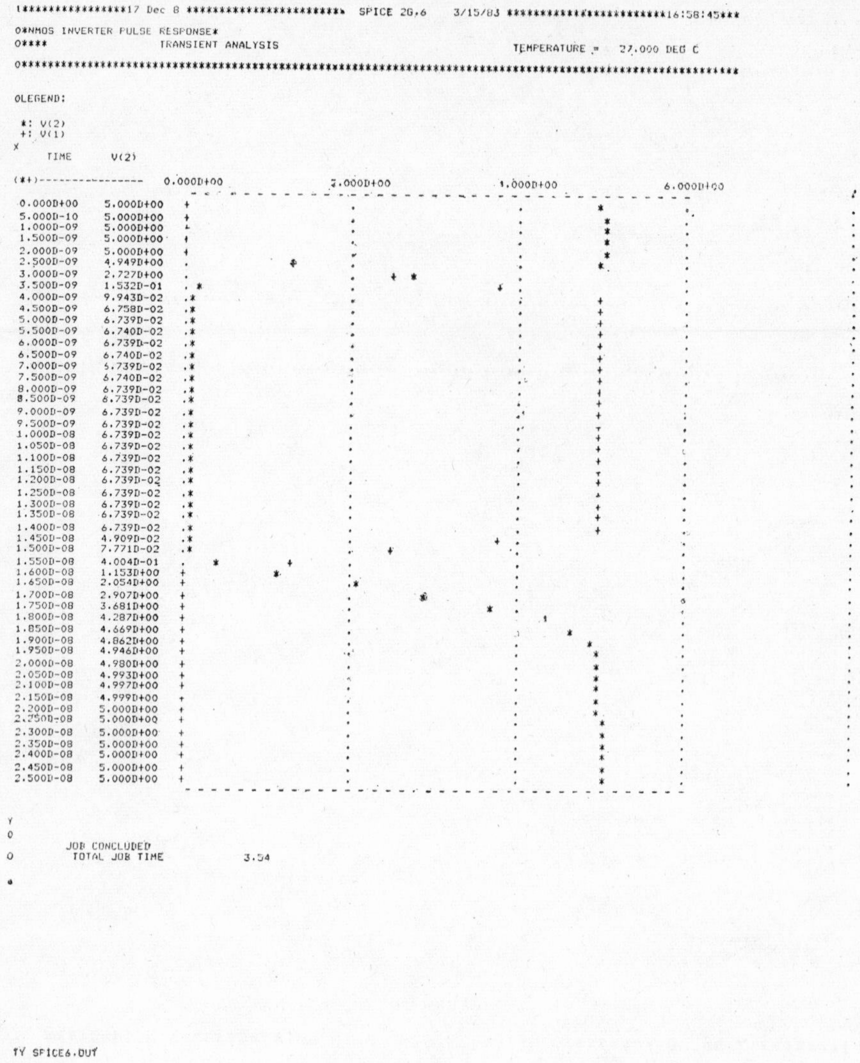

```
1*****************17 Dec 8 **********************  SPICE 2G.6    3/15/83 ******************16:58:45***

0*NMOS INVERTER PULSE RESPONSE*
0****                  TRANSIENT ANALYSIS                              TEMPERATURE =  27.000 DEG C
0*************************************************************************************************

0LEGEND:

  *: V(2)
  +: V(1)
X
     TIME      V(2)

(*+)---------------  0.000D+00          7.000D+00          1.000D+00          6.000D+00
   0.000D+00   5.000D+00    +                     .                    *                  .
   5.000D-10   5.000D+00    +                     .                    *                  .
   1.000D-09   5.000D+00    +                     .                    *                  .
   1.500D-09   5.000D+00    +                     .                    *                  .
   2.000D-09   5.000D+00    +                     .                    *                  .
   2.500D-09   4.949D+00    +                     .                    *.                 .
   3.000D-09   2.727D+00    .                     . *  +               *.                 .
   3.500D-09   1.532D-01    . *                   .          +          .                 .
   4.000D-09   9.943D-02    .*                    .                    +                  .
   4.500D-09   6.758D-02    .*                    .                    +                  .
   5.000D-09   6.739D-02    .*                    .                    +                  .
   5.500D-09   6.740D-02    .*                    .                    +                  .
   6.000D-09   6.739D-02    .*                    .                    +                  .
   6.500D-09   6.740D-02    .*                    .                    +                  .
   7.000D-09   5.739D-02    .*                    .                    +                  .
   7.500D-09   6.740D-02    .*                    .                    +                  .
   8.000D-09   6.739D-02    .*                    .                    +.                 .
   8.500D-09   6.739D-02    .*                    .                    +                  .
   9.000D-09   6.739D-02    .*                    .                    +                  .
   9.500D-09   6.739D-02    .*                    .                    +                  .
   1.000D-08   6.739D-02    .*                    .                    +.                 .
   1.050D-08   6.739D-02    .*                    .                    +                  .
   1.100D-08   6.739D-02    .*                    .                    +                  c
   1.150D-08   6.739D-02    .*                    .                    +                  .
   1.200D-08   6.739D-02    .*                    .                    +                  .
   1.250D-08   6.739D-02    .*                    .                    A                  .
   1.300D-08   6.739D-02    .*                    .                    +                  .
   1.350D-08   6.739D-02    .*                    .                    +                  .
   1.400D-08   6.739D-02    .*                    .                    +                  .
   1.450D-08   4.909D-02    .*                    .          +          +                 .
   1.500D-08   7.771D-02    .*         *          +         +                            .
   1.550D-08   4.004D-01    .        *          *.                                        .
   1.600D-08   1.153D+00    +              *.                                             .
   1.650D-08   2.054D+00    +                          *.                                 .
   1.700D-08   2.907D+00    +                              *                     o        .
   1.750D-08   3.681D+00    +                                    *                        .
   1.800D-08   4.287D+00    +                                        *                    .
   1.850D-08   4.669D+00    +                                          .    *             .
   1.900D-08   4.862D+00    +                     .                         *             .
   1.950D-08   4.946D+00    +                     .                          *            .
   2.000D-08   4.980D+00    +                     .                          *            .
   2.050D-08   4.993D+00    +                     .                          *            .
   2.100D-08   4.997D+00    +                     .                          *            .
   2.150D-08   4.999D+00    +                     .                          *            .
   2.200D-08   5.000D+00    +                     .                          *            .
   2.250D-08   5.000D+00    +                     .                          *            .
   2.300D-08   5.000D+00    +                     .                          *            .
   2.350D-08   5.000D+00    +                     .                          *            .
   2.400D-08   5.000D+00    +                     .                          *            .
   2.450D-08   5.000D+00    +                     .                          *            .
   2.500D-08   5.000D+00    +                     .                          *            .

Y
0
0       JOB CONCLUDED
           TOTAL JOB TIME      3.54
*
```

TY SPICE6.OUT

(d)

Fig. 7.7 Same plot as Fig. 7.6 but with length of depletion mode transistor reduced to half.

No example has been given for small signal analysis here. This is done in the following section after a macro-model for an opamp is given.

7.3 Simulation of OPAMP (Operational Amplifier) Circuits

An opamp is very standard building block for analog circuits. It is therefore very important to be able to simulate circuits containing opamps. Such circuits could be made up of several discrete *IC* opamps or could be just one *IC* containing many opamps. A straightforward way to simulate opamp circuits is to replace every opamp with the actual circuit. A typical BJT opamp (like the 741) contains upwards of 20 transistors. For effective simulation, each transistor has to be replaced by its Gummel Poon model with its 30-odd parameters. A circuit with a few opamps would have more than 100 transistors. This is close to the limit for most programs and computer time required may run into hours. An added difficulty is that the Gummel-Poon parameters for the transistors in a commercial *IC*, are not easy to come by.

An alternative approach, which is discussed here, is to use a macromodel. Such a model is an accurate model as far as the terminal characteristics are concerned. However, it may look totally different from the actual device inside. Naturally, efforts would be made to make the macromodel simpler. This is often possible with the use of idealised elements. For example, a group of five transistors may be used to realise a controlled source. In the macromodel, the group of five highly non-linear transistors can be replaced by one linear controlled source ! The macromodel discussed here is for BJT opamps. But it is easily extended to opamps with FET front ends and CMOS opamps. The model is based on Ref. [4]. It has just two non-linear devices which are BJTs with relatively few non-linear elements in their models.

Figure 7.8 shows the basic opamp macromodel. Values for the various elements have to be chosen appropriately to represent the characteristics of the opamp. A few parameters can be chosen arbitrarily as the macromodel terminal characteristics are not sensitive to values of these parameters.

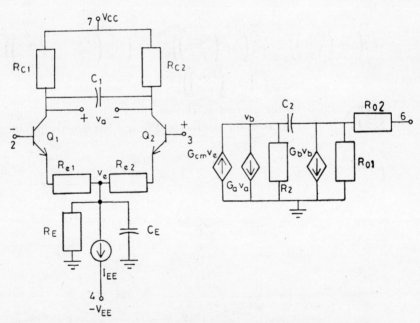

Fig. 7.8 Opamp macromodel without current and voltage limiting features. Terminals marked are corresponding to 741.

Actual opamp characteristics like, say, for the 741 are found from data books. Among these are open loop low frequency gain, slew rate, open loop 3 dB point, input and output impedance, CMRR, etc. The reader is expected to be familiar with these and may refer to a basic book like Ref. [5].

The two input transistors Q_1 and Q_2 in Fig. 7.8 are 'ideal'. All capacitor values are zero and the transistors can be represented by the DC Ebers-Moll model having two non-linear resistors and two CCCS'. The only parameters to be specified for Q_1 and Q_2 are I_{S1}, I_{S2}, β_{F1} and β_{F2} (β_S here are forward β_S). Of these β_{F1} and β_{F2} are chosen to be slightly different to account for the offset in bias currents and I_{S1} and I_{S2} are chosen to be slightly different to account for the voltage input offset. If I_B is the average input bias current and I_{BOS} the offset in bias current

$$I_{B1} = I_B + \frac{I_{BOS}}{2} \qquad I_{B2} = I_B - \frac{I_{BOS}}{2}$$

Taking the quiescent collector currents $I_{C1} = I_{C2} = I_C$, we get

$$\beta_{F1} = \frac{I_C}{I_{B1}}, \qquad \beta_{F2} = \frac{I_C}{I_{B2}}$$

I_C is determined later. It may be noted that while the bias current itself may be large (hundreds of nAs), the offset is kept to a low value (less than a few nAs) in a good opamp.

In order to maintain $I_{C1} = I_{C2}$, there must be a small input voltage offset given by the difference between V_{BE1} and V_{BE2}. I_{S1} is chosen arbitrarily as 8×10^{-16} A. I_{S2} is determined from the specification V_{OS} for input voltage offset

$$I_{C1} = I_{C2} \text{ implies}$$

$$I_{S1}\left(\exp\left(\frac{V_{BE1}}{V_T}\right)\right) = I_{S2}\left(\exp\left(\frac{V_{BE2}}{V_T}\right)\right) = I_{S2}\left(\exp\left(\frac{(V_{BE1} - V_{OS})}{V_T}\right)\right)$$

or

$$I_{S2} = I_{S1}\left(\exp\left(\frac{V_{OS}}{V_T}\right)\right)$$

The quiescent collector current is chosen from the positive going slew rate S_R^+. In Fig. 7.9, the relevant portion of Fig. 7.8 has been reproduced.

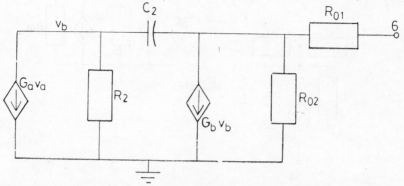

Fig. 7.9 Relevant part of opamp macromodel to determine quiescent collector current.

The maximum rate of rise possible at the output is when the entire current $G_a v_a$ flows through C_2. This condition corresponds to S_R^+ when v_a is maximum. Therefore

$$S_R^+ = \left(\frac{dv_6}{dt}\right)_{max} = \frac{G_a v_{amax}}{C_2} = \frac{G_a 2I_c R_c}{C_2}$$

where $$R_c = R_{c1} = R_{e2}$$

For convenience $G_a R_c$ is chosen to be unity. This gives

$$I_c = \frac{S_R^+ C_2}{2}$$

The capacitance C_2 itself represents the actual capacitance present in the opamp to introduce a dominant pole. Its value is 30 pF for the 741 opamp.

The collector resistance R_c is determined from the open loop unity gain frequency. Using the Miller equivalent shown in Fig. 7.10

$$f_{3dB} = \frac{1}{2\pi R_2 C_{in}} = \frac{1}{2\pi R_2 G_b R_{02} C_2}$$

where f_{3dB} represents the open loop 3 dB frequency. If the open loop low frequency gain is a_{vDO},

$$a_{vD} = a_{vDO} \Big/ \left(1 + j\frac{f}{f_{3dB}}\right)$$

The unity gain frequency (for $a_{vDO} \gg 1$) $f_0 = a_{vDO} f_{3dB}$ and a_{vDO} is given by

$$a_{vDO} = G_a R_2 G_b R_{02}$$

as the input stage is designed to have unity differential gain. Therefore

$$f_0 = \frac{1}{2\pi C_2 R_C}$$

using $$G_a R_C = 1$$

$$R_C = \frac{1}{2\pi f_0 C_2}$$

The quantity f_0 also represents the gain bandwidth product of the opamp.

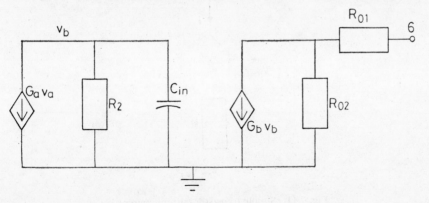

Fig. 7.10   Miller equivalent corresponding to Fig. 7.9.

The emitter degeneracy resistors R_{e1} and R_{e2} are chosen to make the differential gain of the input stage $= 1$. It can be shown that [6]

$$\frac{v_a}{v_{in}} = \frac{v_a}{v_+ - v_-} = \frac{\beta_{F1}R_{C1} + \beta_{F2}R_{C2}}{h_{ie1} + (\beta_{F1} + 1)R_{e1} + h_{ie2} + (\beta_{F2} + 1)\ R_{e2}}$$

Equating the above expression to 1 and putting $R_{C1} = R_{C2}$,

$$R_{e1} = R_{e2} = R_e \quad \text{and} \quad h_{ie} = \beta_F\frac{V_T}{I_C} = \frac{\beta_F}{g_m},$$

we get

$$1 = \frac{R_C(\beta_{F1} + \beta_{F2})}{R_e(\beta_{F1} + \beta_{F2} + 2) + h_{ie1} + h_{ie2}}$$

$$R_e = \frac{\beta_{F1} + \beta_{F2}}{(\beta_{F1} + \beta_{F2} + 2)}\left(R_C - \frac{V_T}{I_C}\right)$$

The current source I_{EE} is given by

$$I_{EE} = I_C\left[\frac{(\beta_{F1} + 1)}{\beta_{F1}} + \frac{(\beta_{F2} + 1)}{\beta_{F2}}\right]$$

The resistor R_E in parallel with I_{EE} represents the common mode input impedance. It results from the output impedance of the circuit used to realise the current source I_{EE}. Basically, the circuit would be as shown in Fig. 7.11 with the current given by

$$I(= I_{EE}) = \frac{V_{in} - 0.7 + V_{EE}}{R}$$

The output impedance is just the variation of collector current as a function of V_{CE} for the transistor at constant I_B. As discussed in Section 6.3, this can be represented by the forward Early Voltage V_{AF}. Therefore,

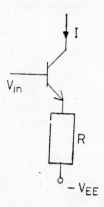

Fig. 7.11   Simple transistor current source.

$$R_E \simeq \frac{V_{AF}}{I_{EE}} \quad \text{with} \quad V_{AF} = 200 \text{ V}$$

The choice of R_2 is not critical. In the active region it is the product $G_b R_2$ that must have a specific value. For proper transient response a value $R_2 = 100$ K is found suitable. The VCCS G_{cm} is used to model the common mode gain. The common mode rejection ratio CMRR is given by

$$\text{CMRR} = \frac{G_a}{G_{cm}} \quad \text{or} \quad G_{cm} = \frac{1}{(\text{CMRR}) \times R_c}$$

At low frequencies the output impedance is given by

$$R_{\text{OUT}} = R_{01} + R_{02}$$

At higher frequencies it is complex. At very high frequencies it is real again and given by

$$R_{\text{OUT}_{AC}} = R_{01}$$

R_{01} and R_{02} are given by values of R_{OUT} and $R_{\text{OUT}_{AC}}$. Actual variation of Z_{OUT} with frequency can be found by taking the Thevenin equivalent between point 6 and ground (G_{cm} is ignored) using the circuit of Fig. 7.9. This gives

$$\bar{V}_{OC} = \frac{-G_a v_a (jwC_2 - G_b)}{jwC_2(G_{02} + G_b + G_2) + G_2 G_{02}}$$

$$\bar{I}_{SC} = \frac{1}{R_{01}} \frac{(-) G_a v_a (jwC_2 - G_b)}{jwC_2(G_{02} + G_{01} + G_2 + G_b) + G_2(G_{02} + G_{01})}$$

where,

$$G_{01} = \frac{1}{R_{01}}, \quad G_{02} = \frac{1}{R_{02}}, \quad G_2 = \frac{1}{R_2}$$

and $\bar{V}_{OC}$ and $\bar{I}_{SC}$ are phasors

$$Z_{\text{OUT}} = Z_{\text{Thevenin}} = \frac{\bar{V}_{OC}}{\bar{I}_{SC}}$$

$$Z_{\text{OUT}} = R_{01}\left[1 + \frac{jwC_2 G_{01} + G_2 G_{01}}{jwC_2[G_{02} + G_b + G_2] + G_2 G_{02}}\right]$$

At $\omega = 0$, we get $Z_{\text{OUT}} = R_{01} + R_{02}$ and at $\omega = \infty$, we get $Z_{\text{OUT}} = R_{01}$ as $G_b \gg G_{02} \gg G_2$. At a frequency

$$\omega = \frac{1}{C_2 R_{02} G_b R_2}$$

the impedance Z_{OUT} is given by

$$R_{01} + \frac{R_{02}(i-j)}{2}$$

G_b can now be calculated from the low frequency open loop differential gain as

$$G_b = \frac{a_{vDO}R_C}{R_2 R_{O2}}$$

The capacitor C_1 is used to take care of higher order poles. Considering the dominant pole alone, the open loop gain is given by

$$a_{vD} = \frac{a_{vDO}}{(1 + jf/f_{3dB})}$$

At f_{0dB} or f_0 the phase would be $\phi = -\tan^{-1} f_0/f_{3dB}$ which is approximately $-90°$. Now assume another pole at a frequency $f_2(f_2 \gg f_{3dB})$ so that

$$a_{vD} = \frac{a_{vDO}}{\left(1 + \dfrac{jf}{f_{3dB}}\right)\left(1 + \dfrac{jf}{f_2}\right)}$$

The phase of ϕ at f_0 will now be

$$\phi = -90° - \tan^{-1}\left(\frac{f}{f_2}\right)$$

In other words, an excess phase $(-\tan^{-1} f/f_2)$ is introduced. This has the effect of reducing the phase margin by that amount. If ϕ become $-180°$ at f_0, then the opamp would become unstable. From the specified value of this excess phase, C_1 can be calculated. C_1 introduces a pole at a frequency equal to $\dfrac{1}{(2\pi 2 R_C C_1)}$ as can be found from an analysis of the input differential stage. This analysis can be done by replacing each transistor with the h-parameter model. C_1 effectively appears across the series combination of R_{C1} and R_{C2}. We now have

Magnitude of excess phase $|\Delta\phi|$ at $f_0 = \tan^{-1}(f_0 2\pi 2 R_C C_1)$

or,

$$C_1 = \frac{(\tan|\Delta\phi|)}{4\pi f_0 R_C} = \frac{C_2 \tan|\Delta\phi|}{2}$$

The capacitor C_E at the emitter has been put in to model the differential slew rate of the opamp. For an opamp with npn transistors at its input differential stage, the slew rate for positive going output S_R^+ is greater than that for a negative going output S_R^-. For pnp input transistors, the opposite is true. This is because both C_E and C_2 have to be charged during conditions when S_R^- applies.

In order to understand this, consider an opamp in the voltage follower mode as shown in Fig. 7.12. Let a negative step be given at the $+$ input. This tends to turn Q_1 on and Q_2 off. The emitter capacitance C_E effectively appears across the output. Q_1 is operating in the emitter follower mode and $v_e \simeq v_0$. In response to a negative step, v_0 and v_e go negative and C_E is discharged with a current i_{CE} flowing with the polarity shown. The maximum

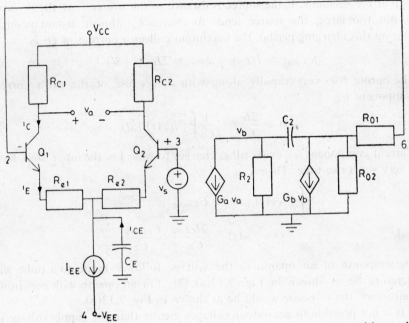

Fig. 7.12 Opamp connected in the voltage follower mode to study positive and negative slew rates. For a positive step at '+' input, direction of i_{CE} would be reversed.

value for the current $i_E(\simeq i_C)$ is $(I_{EE} - i_{CE})$ which in turn means

$$v_{a\,max} = (I_{EE} - i_{CE})R_C.$$

As the product $G_a R_C = 1$, we can write

$$\left(\frac{dv_0}{dt}\right)_{max} \simeq \left(\frac{dv_e}{dt}\right)_{max} \simeq \frac{(I_{EE} - i_{CE})}{C_2}$$

$$I_{EE} \simeq 2I_C \quad \text{and} \quad i_{CE} = C_E\left|\frac{dv_e}{dt}\right|$$

$$\left|\frac{dv_0}{dt}\right|_{max} = \frac{1}{C_2}\left[2I_C - C_E\left|\frac{dv_0}{dt}\right|_{max}\right]$$

or

$$\left(\frac{dv_0}{dt}\right)_{max} = S_R^- = \frac{2I_C}{C_2 + C_E}$$

and

$$C_E = \frac{2I_C}{S_R^-} - C_2$$

When a positive step is given at the + input in the voltage follower mode, Q_1 tends to go off disconnecting the input stage from the output. The capacitor C_2 alone needs to be charged and therefore S_R^+ is greater than S_R^-. Actually the situation is a little more complex even for a positive step at the $+$ input. If the source resistance of the source v_s at the $+$ input is not too large, the source tends to charge C_E almost instantaneously. During this charging period, the maximum collector current of Q_2 is

$$i_{C2\,max} = I_{EE} + |\,i_{CE}\,| \simeq 2I_C + |\,i_{CE}\,|$$

The output rises very rapidly along with v_s because of the extra current component i_{CE}

$$v_0(t) = \frac{2I_C}{C_2}t + \frac{1}{C_2}\int |\,i_{CE}(\tau)\,|\,d\tau$$

Current component i_{CE} exists till v_e charges to v_{step}. Let the input step have a very short rise time. Then

$$\int |\,i_{CE}(\tau)\,|\,d\tau = C_E v_{step}$$

and

$$v_0(t) = \frac{2I_C t}{C_2} + \frac{C_E v_{step}}{C_2}$$

The response of an opamp in the voltage follower mode to a pulse will therefore be as shown in Fig. 7.13(a) [7]. For an opamp with *pnp* input transistors, the response would be as shown in Fig. 7.13(b).

It is not possible to get output voltages greater than the supply voltage in magnitude. This voltage limiting feature is incorporated in the macromodel as shown in Fig. 7.14. If V_{OUT}^+ is the maximum positive output voltage and V_{OUT}^- the maximum output voltage

$$V_C = V_{CC} - V_{OUT}^+ + V_{D3}$$
$$V_E = |\,V_{EE}\,| - |\,V_{OUT}\,|^- + V_{D4}$$

V_{D3} and V_{D4}, being diode drops, are about 0.7 V. Their reverse saturation currents can be chosen rather arbitrarily.

An opamp also has short circuit protection i.e. even when short circuited the output current does not exceed about 30 mA in either direction. This is simulated by the circuit segment shown in Fig. 7.15 which has two diodes and a VCVS. Whenever the current I_0 is such that $|\,I_0 R_{01}\,|$ exceeds 0.6 V, either $D1$ or $D2$ will begin to conduct and divert the extra current. If I_{SC} represents the maximum short circuit current (note I_{SC}^+ and I_{SC}^- may be slightly different), then $I_{SC}R_{01}$ should be 0.6 to 0.7 V. However R_{01} has been determined from output impedance considerations. If $I_{SC}R_{01}$ is not around 0.7 V, somewhat unrealistic values for the reverse saturation currents of D_1 and D_2 may be necessary. Table 7.1 gives opamp specifications and macro-model values for a LM741 opamp. The maximum current available from the VCCS G_b is $2I_C R_2 G_b$. From values given in Table 7.1 is thus about 500 A. It must be remembered that such a large current can exist only in the macro model and not in the opamp itself. This large current must be diverted by $D1$ or $D2$. This gives the reverse saturation current for $D1$ and $D2$.

$$500\text{ A} = I_{SD1}\,\exp\left(\frac{I_{SC}R_{01}}{V_T}\right)$$

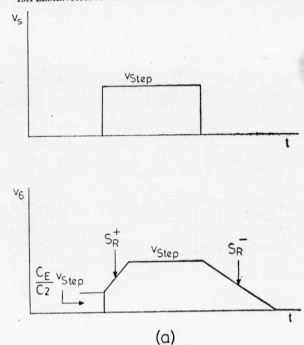

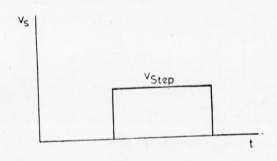

(a)

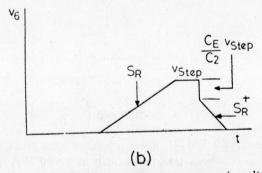

(b)

Fig. 7.13   Response to a pulse for an opamp in voltage
follower mode with (a) *npn* transistors for
input differential stage (b) *pnp* transistors
for input differential stage.

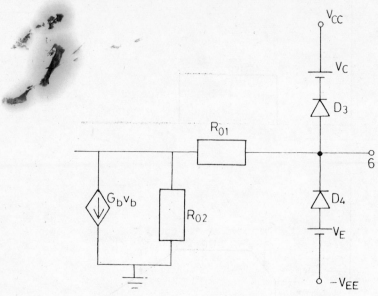

Fig. 7.14 Voltage limiting features in macromodel.

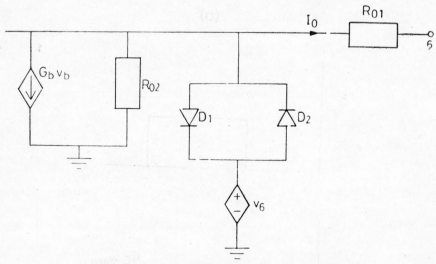

Fig. 7.15 Current limiting features in macromodel.

or

$$I_{SD1} = I_{SD2} = 500 \ \exp \left(\frac{-I_{SC} R_{01}}{V_T} \right)$$

In the specifications for LM741, $R_{OUT_{AC}}$ has not been given. So R_{01} is not known. Here I_{SD1} has been taken arbitrarily as 8×10^{-16} A and R_{01} chosen from the equation above.

Finally the total dissipation in the macromodel is made equal to that in the actual opamp by means of a resistor R_P between V_{CC} and $-V_{EE}$. If P_d

Table 7.1   OPAMP Specifications and Macromodel Parameters for LM741

OPAMP Specifications	Macromodel Parameter
(from data sheets)	
$(V_{CC} = -V_{EE} = 15$ V$)$	$(V_{CC} = -V_{EE} = 15$ V$)$
$(C_2 = 30$ pF$)$	(Temp. 300°K)
$S_R^+ = 0.67$ V/μ sec	$I_{S1} = 8 \times 10^{-16}$ A
$S_R^- = 0.62$ V/μ sec	$I_{SD3} = 8 \times 10^{-16}$ A
$I_B = 80$ nA	$R_2 = 100$ K
$I_{BOS} = 20$ nA	$C_2 = 30$ pF
$V_{OS} = 1$ mV	$C_E = 2.41$ pF
$a_{vDO} = 2 \times 10^5$	$\beta_{F1} = 111.67$
$a_{vD}(1$ KHz$) = 10^3$	$\beta_{F2} = 143.57$
$\Delta\phi = 20°$	$I_{EE} = 20.26$ μA
CMRR = 90 dB	$R_E = 4.872$ Mohms
$R_{OUT} = 75$	$I_{S2} = 6.309 \times 10^{-16}$ A
$R_{OUT_{AC}} = -$	$R_C = 5305$ Ohms
$I_{SC}^+ = 25$ mA $= I_{SC}^-$	$R_e = 2712$ ohms
$V_{OUT}^+ = 14.2$ V	$C_1 = 5.46$ pF
$V_{OUT}^- = -12.7$ V	$R_p = 18$ kΩ
$P_d = 50$ mW	$G_a = 188.6$ μ mhos
	$G_{CM} = 6.28$ n mhos
	$R_{O_1} = 32.13$ ohms
	$R_{O_2} = 42.87$ ohms
	$G_b = 247.49$ mhos
	$I_{SD1} = 8 \times 10^{-16}$ A
	$V_C = 1.803$ V
	$V_E = 2.303$ V

is the DC power dissipation of the opamp then,

$$R_p = \frac{(V_{CC} + V_{EE})^2}{P_d - V_{CC}(2I_c) - V_{EE}I_{EE}}$$

The complete opamp macromodel is as shown in Fig. 7.16. Table 7.1 gives the opamp specifications from the data book and the calculated opamp parameters for the LM741 opamp.

As an example, a biquad bandpass filter was analysed using the macromodel. The current and voltage limiting features were left out. The filter is shown in Fig. 7.17. For ideal opamps, the design equations are

Resonance frequency $f_r = \dfrac{1}{2\pi C(R_2 R_3)^{1/2}}$

Bandwidth $BW = \dfrac{f_r}{Q} = \dfrac{1}{2\pi R_1 C}$

Gain at $f_r = \dfrac{R_1}{R_4} = A_r$

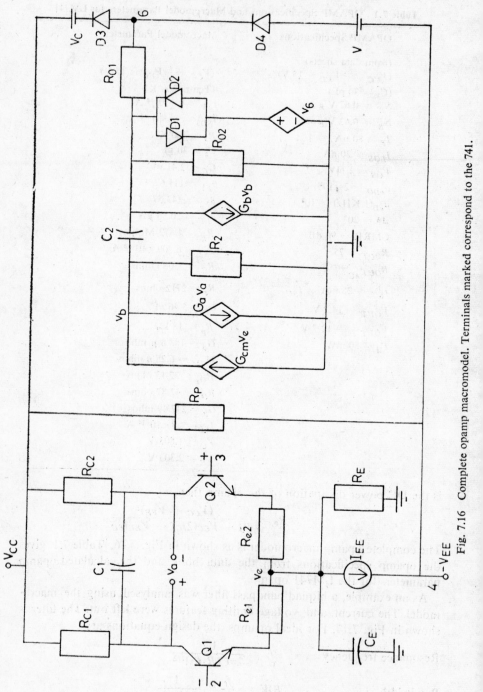

Fig. 7.16   Complete opamp macromodel. Terminals marked correspond to the 741.

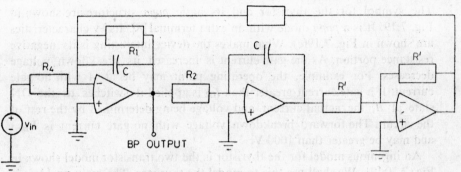

Fig. 7.17 Biquad bandpass filter.

For non-ideal opamps, the realised Q is higher because of the finite gain bandwidth product f_0 of the opamp.

$$Q_{realised} = \frac{Q_{design}}{1 - \frac{4 f_r Q_{design}}{f_0}}$$

Figure 7.18 gives the SPICE ouput for a filter with $f_r = 2.5$ kHz, $BW = 50$ Hz ($Q_{design} = 50$), $A_r = 1$, $R_2 = R_3$ and $C = 0.1$ μF. The realised Q is much higher and is approximately equal to the value given by the equation above. Using parameter values in Table 7.1,

$$Q_{realised} = \frac{50}{1 - \frac{4 \times 2.5 \times 10^3 \times 50}{10^6}} = 100$$

7.4 Simulation of Thyristor Circuits

A macromodel for an IC opamp was described in the last section. This macromodel greatly simplified the simulation of opamp circuits. A different kind of model is developed in this section. A thyristor is a semiconductor device frequently used in power electronics. However, no model exists for it in programs like SPICE. We show how a model for a thyristor can be constructed from models which are built-in in SPICE. Usually it is a lot easier to do this than to try and incorporate one's own sub-routine in a large and complex program like SPICE.

A thyristor or silicon controlled rectifier (SCR) is like a diode in function in that it conducts only in one direction. When ON, it can carry a large current (upto thousands of amperes) with less than a volt drop across it. When OFF, it can have a large voltage (upto thousands of volts) with milli-amperes of current flowing through it. In either case the dissipation in the device is small and dissipation occurs mainly during the switching periods. Like a diode, it has an anode and a cathode. In addition, there is a third terminal called gate. Even when forward biased, the thyristor remains OFF till a gate current pulse is given. Reverse bias always turns off the diode.

The symbol for the thyristor and its basic *pnpn* structure are shown in Fig. 7.19. It is a *pnpn* diode with an extra terminal [8]. Its i-v characteristics are shown in Fig. 7.19(c). What makes the device interesting is its negative resistance portion. As the gate current is increased its breakdown voltage decreases. For example, the operating point may be at A with no gate current. If a gate current greater than I_{G2} is applied it switches to the ON state at B, the actual current, and voltage being determined by the rest of the circuit. The forward breakdown voltage with no gate current is V_{BFO} and may be greater than 1000 V.

An ingenious model for the thyristor is the two transistor model shown in Fig. 7.20 [9]. We shall use this to model the thyristor. The main problem is to appropriately select the Gummel-poon parameters of Q_p and Q_n. The various p and n layers are sandwiched as shown in Fig. 7.19(b). Both Q_n and Q_p are therefore lateral transistors as opposed to the normal vertical transistors. They have low values of β_F. Let the thyristor represented by the two transistor model be forward biased with $I_G = 0$ and V_{AK} a fairly large value, say, 100 V. Assume V_{BFO} to be greater than 100 V. Then the thyristor

```
1********18 Dec 8 ********  SPICE 2G.6   3/15/83 *******16:48:53******

0*FREQUENCY RESPONSE OF BIQUAD BANDPASS FILTER*

0****      INPUT LISTING                    TEMPERATURE =   27.000 DEG C

0******************************************************************************


        VCC 7 0 15V
        VEE 0 4 15V
        V1 1 0 AC 1V
        R1 2 3 31.8K
        R2 3 5 630
        R3 9 3 630
        R4 1 2 31.8K
        R5 3 8 10K
        R6 8 9 10K
        C1 2 3 1.8NF
        C2 5 6 10CNF
        .SUBCKT OPAMP 2 3 6 7 4
        Q1 9 2 5 BJ1
        Q2 10 3 13 BJ2
        IEE 8 4 27.30A
        RC1 7 9 4.35K
        RC2 7 10 4.35K
        RE1 5 8 2.4K
        RE2 13 8 2.4K
        CE 8 0 7.5PF
        C2 11 12 30PF
        R2 11 0 100K
        E02 12 0 4V0
        RO1 12 6 77
        G1 11 0 9 10 0.25MMHOS
        G2 12 0 11 0 37.1
        .MODEL BJ1 NPN BF=52.67 IS=8.0E-16
        .MODEL BJ2 NPN BF=52.79 IS=8.09E-16
        .ENDS OPAMP
        X1 2 0 3 7 4 OPAMP
        X2 5 0 6 7 4 OPAMP
        X3 8 0 9 7 4 OPAMP
        .AC LIN 50 2KHZ 3KHZ
        .PLOT AC VM(3)
        .END
```

(a)

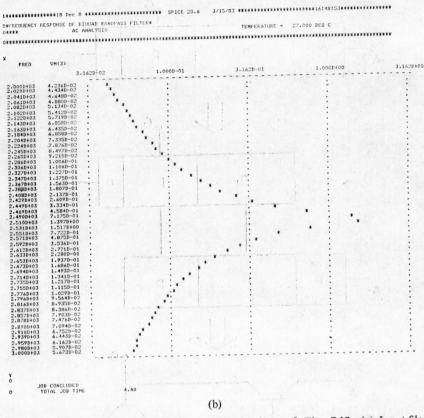

(b)

Fig. 7.18 SPICE frequency response of Biquad filter of Fig. 7.17. (a) Input file
(b) Plot of response.

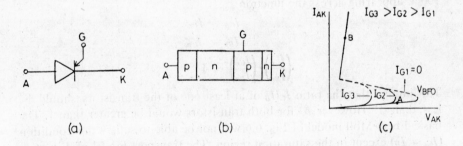

(a)   (b)   (c)

Fig. 7.19 (a) Symbol for a thyristor. (b) Basic p-n-p-n structure showing anode (A),
cathode (K) and gate (G). (c) i-v characteristic of a thyristor for different
gate currents

should be OFF. The E-B junctions of both Q_n and Q_p would be forward
biased with the bias slightly less than the cutin voltage. The common B-C
junction of the two transistors would be reverse biased with most of the

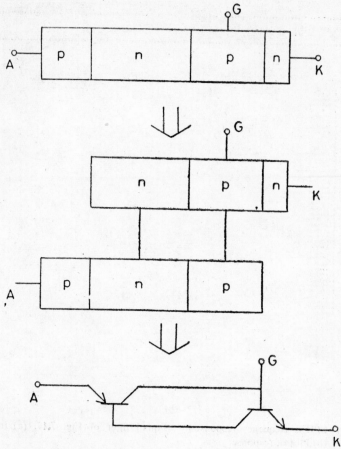

Fig. 7.20   Two transistor model for a thyristor.

100 V appearing across the junction.

$$I_{Cp} = I_{Bn}$$

$$I_{Bp} = I_{Cn}$$

$$\left(\frac{I_{Cp}}{I_{Bp}}\right) \cdot \left(\frac{I_{Cn}}{I_{Bn}}\right) = 1$$

This implies that the ratio I_C/I_B of at least one of the transistors should be less than 1. However β_F for both transistors would be greater than 1. The basic Ebers-Moll model of Fig. 6.6 will not be able to satisfy this condition ($I_C \Leftarrow I_B$) except in the saturation region. The transport model of Fig. 6.8, however, can satisfy this condition because of the presence of space charge recombination current. This current is modelled by the emitter-base diode having the reverse saturation current I_{SE} with the emission constant η_E approximately equal to 2. The condition $I_C = I_B$ is met when the collector-base junction is reverse biased and the emitter-base junction has a low value of forward bias. For low values of forward bias the space charge recombination current dominates in the emitter-base junction as I_{SE} is 100 to 1000 times I_S. (Incidentally, space charge recombination current is negligible in germanium. A *pnpn* structure with germanium cannot therefore have an OFF state in forward bias. Hence there are no germanium controlled rectifiers.) Space charge recombination currents have to be included using the parameters I_{SE} and η_E in SPICE before the OFF state of the thyristor can be modelled.

Figure 7.21 shows typical dimensions and doping profiles for a thyristor [10]. From these the following approximate values for the Gummel-Poon parameters of Q_P and Q_n can be calculated. A current rating of 5 A is assumed.

<table>
<tr><td>

pnp transistor Q_p

Area $= 0.1$ cm^2

$I_S = 10^{-11}$ A

$I_{SC} = I_{SE} = 1000 \, I_S$

$\eta_E = 2$

$\beta_F = 2$

$\beta_R = 2$

$V_{AF} = 1000$ V

$C_{JC} = C_{JE} = 10$ pF

</td><td>

npn transistor Q_n

Area $= 0.1$ cm^2

$I_S = 10^{-14}$ A

$I_{SC} = I_{SE} = 100 \, I_S$

$\eta_E = 2$

$\beta_F = 5$

$\beta_R = 0.1$

$V_{AF} = 200$ V

$C_{JC} = C_{JE} = 10$ pF

</td></tr>
</table>

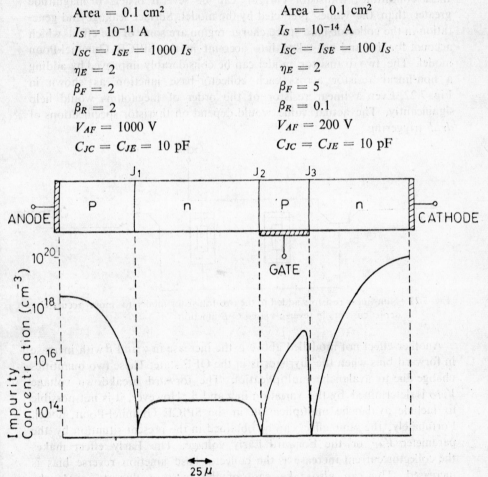

Fig. 7.21 Typical dimensions and doping profiles for a thyristor. [10]

When these values are used in the SPICE Gummel-Poon model the following thyristor specifications are poorly matched.

(1) The model predicts a forward bias OFF state current of the order of nA. The actual current is typically 15 mA.
(2) The minimum gate current for triggering at 100 V forward bias is 0.2 μA, while actual values are 1 to 10 mA.
(3) The value of dv/dt for breakdown or triggering in the absence of gate currents is 0.03 V/microsec from the model while actual values are about 100 V/microsec.

These problems arise due to the inadequacy of the Gummel-Poon model for a transistor which is OFF. For a thyristor in the OFF state, the emitter-base junctions are forward biased to a value slightly less than the cutin voltage and the collector-base junctions are heavily reverse biased. Under these conditions, transistor currents can be several orders of magnitude greater than the values predicted by the model. Surface leakage and generation in the collector-base space charge region are some of the effects which account for this and not taken into account in the SPICE Gummel-Poon model. The two transistor model can be considerably improved by adding a non-linear resistor across each collector-base junction as shown in Fig. 7.22. Even a linear resistor of the order of megaohms would help significantly. The actual value would depend on thyristor specifications of dv/dt triggering.

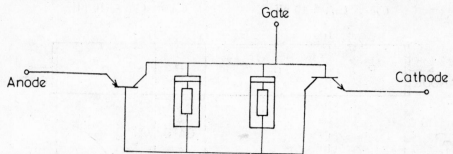

Fig. 7.22   Non-linear resistors added to the two-transistor model to more accurately depict currents in reverse-biased C–B junction.

Another effect not modelled above is the increase in α and β with increase in forward bias when the thyristor is in the OFF state. These two quantities change due to avalanche multiplication. The forward breakdown voltage V_{BFO} is determined by the variation in α and β. However, it is not possible to include avalanche multiplication in the SPICE Gummel-Poon model. Fortunately, the same effect can be obtained in the present situation by the parameter V_{AF} or the Forward Early voltage. The Early effect makes the collector current increase as the collector-base junction reverse bias is increased. This can also take care of the increase due to avalanche multiplication.

Many thyristors have shorted emitters, i.e. the cathode and gate are shorted [10]. This improves the dv/dt rating. In these thyristors the gate terminal is physically located far from the short. Such a thyristor can be modelled as an unshorted thyristor (near the gate) in parallel with a shorted emitter thyristor (nearer the gate-cathode short). Applying the two transistor model to each thyristor we end up with the circuit of Fig. 7.23. The resistor r models the resistance from the gate-metal contact to the cathode-metal contact.

The following parameters were chosen for the various elements in Fig. 7.23.

Q_3 and Q_1:

$I_S = 10^{-11}$ A

$I_{SE} = 1000\ I_S$

$\beta_F = 2$

$\beta_R = 1$

$C_{JC} = C_{JE} = 10$ pF

$TF = 10\ n$ sec; $TR = 20\ n$ sec

Q_2:

$I_S = 10^{-14}$ A

$I_{SE} = I_{SC} = 100\ I_S$

$\beta_F = 5$

$\beta_R = 0.1$

$C_{JC} = C_{JE} = 10$ pF

$TF = 2\ n$ sec; $TR = 10\ n$ sec

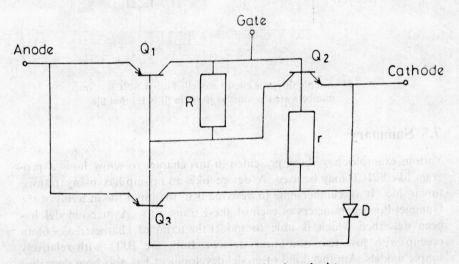

Fig. 7.23   Model for a shorted emitter thyristor.

D:

$$Is = 10^{-14} \text{ A}$$

$$VJ = 0.75 \text{ V}$$

$$r = 500 \ \Omega$$

$$R = 2 \text{ Mega } \Omega$$

The following were the specifications of the thyristor modelled by the above values.

Forward breakdown voltage $V_{BFO} = 2.9$ kV

Minimum gate current for triggering = 2 mA

dv/dt breakdown value = 157 V/microsec

Figure 7.25 shows the SPICE output hwen a gate pulse is given at $\theta = 90°$ for the circuit shown in Fig. 7.24. Node numbers in Fig. 7.24 correspond to the SPICE output in Fig. 7.25.

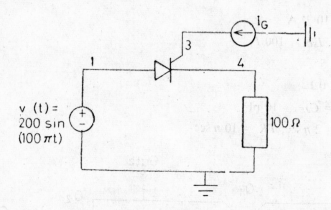

Fig. 7.24 Simple thyristor circuit modelled using SPICE. Note numbers are the same as those in SPICE input file.

7.5 Summary

Various examples have been presented in this chapter to show how a program like SPICE may be used. A device like an opamp has many transistors inside. It is cumbersome to describe it in the input file in terms of the Gummel-Poon parameters of each of these transistors. A macromodel has been described which is able to model the terminal characteristics of an opamp with just two non-linear devices. Both are BJTs with relatively simple models. Another kind of model development has also been described where a thyristor has been modelled using the two transistor model.

```
1::::::: 7 Jul 8 :::::::: SPICE 26.6   3/15/83 ::::::::12:43:06:::::

0:SCR SIMULATION:

0::::   INPUT LISTING            TEMPERATURE =   27.000 DEG C

0::::::::::::::::::::::::::::::::::::::::::::::::::::::::::::::::::::::::::::


    IS 5 3 PULSE(0 5MA 5MS 100US 100US 200US 20MS)
    VZ 5 0 0V
    Q1 4 2 1 MODP
    Q3 3 2 1 MODP
    Q2 2 3 4 MODN
    D1 4 2 DMOD
    R1 3 2 2MEG
    R2 3 4 500
    RL 4 0 100
    .MODEL MODP PNP IS=1.0E-11 ISC=1.0E-08 ISE=1.0E-08 BF=2.0 BR=1.0
    +CJC=10PF CJE=10PF TF=10NS TR=20NS
    .MODEL MODN NPN IS=1.0E-14 ISC=1.0E-12 ISE=1.0E-12 BF=5.00 BR=0.1
    +CJC=10PF CJE=10PF TF=2NS TR=10NS
    .MODEL DMOD D IS=1.0E-14 VJ=0.75V
    VS 1 0 SIN(0 200V 50HZ)
    .TRAN 200US 12MS
    .PRINT TRAN V(1) I(VS) I(VZ) V(4) V(1,4)
    .PLOT TRAN V(1) I(VS) I(VZ) V(1,4)
    .END



1:::::::::::::::: 7 Jul 6 :::::::::::::::::::::::::: SPICE 26.6   3/15/83 ::::::::::::::::::::::::::::::12:43:06::::::::::::::::::

0:SCR SIMULATION:
0::::           TRANSIENT ANALYSIS              TEMPERATURE =   27.000 DEG C

0::::::::::::::::::::::::::::::::::::::::::::::::::::::::::::::::::::::::::::::::::::::::::::::::::::::::::::::::::


0:LEGEND:

$: V(1)
+: I(VS)
=: I(VZ)
$: V(1,4)
I
    TIME      V(1)

($$)-------------- -2.000D+02        -1.000D+02         0.000D+00          1.000D+02         2.000D+02
                   - - - - - - - - - - - - - - - - - - - - - - - - - - - - - - - - - - - - - - - -

(+)-------------- -2.000D+00        -1.000D+00         0.000D+00          1.000D+00         2.000D+00
                   - - - - - - - - - - - - - - - - - - - - - - - - - - - - - - - - - - - - - - - -

(=)-------------- -6.000D-03        -4.000D-03         -2.000D-03         0.000D+00         2.000D-03
                   - - - - - - - - - - - - - - - - - - - - - - - - - - - - - - - - - - - - - - - -

0.000D+00    0.000D+00  .                  .              X            =            .
2.000D-04    1.256D+01  .                  .              + X          =            .
4.000D-04    2.506D+01  .                  .              +   X        =            .
6.000D-04    3.745D+01  .                  .              +     X      =            .
8.000D-04    4.970D+01  .                  .              +       X    =            .
1.000D-03    6.177D+01  .                  .              +         X  =            .
1.200D-03    7.361D+01  .                  .              +           X =           .
1.400D-03    8.513D+01  .                  .              +             X =          .
```

Fig. 7.25   (Contd.)

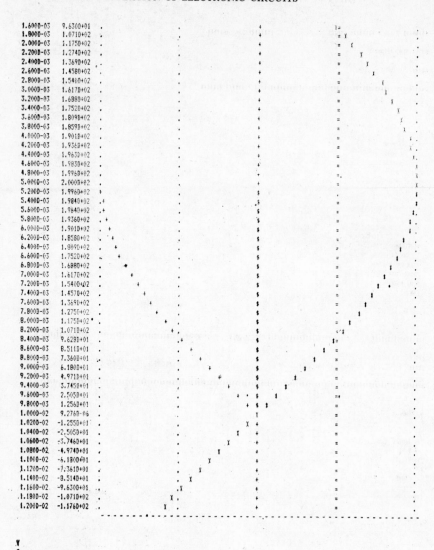

Fig. 7.25 SPICE output for the circuit of Fig. 7.24. Note that the thyristor turns ON just after the gate current pulse at $t = 5$ ms. Thyristor goes OFF when it is reverse biased at $t = 10$ ms.

References

1. L.W. Nagel, SPICE 2: A Computer Program to Simulate Semiconductor Circuits, Ph.D. Thesis, University of California, Berkeley, May 1978.
2. A. Vladimirescu, K. Zhang, A.R. Newton, D.O. Pederson, A.L. Sangiovanni—Vincentelli, SPICE Version 2 G User's Guide, Dept. of Electrical Engineering and Computer Science, Univ. of California, Berkeley, 1981.
3. L.A. Glaser and D.W. Dobberpuhl, The Design and Analysis of VLSI Circuits, pp. 137–139, Addison–Wesley Publishing Co., 1985.

4. G.R. Boyle, B.M. Cohn, D.O. Pederson and J.E. Solomon, Macromodelling of Integrated Circuit Operational Amplifiers, IEEE J. Solid State Ccts, SC-9, pp. 353–363, Dec. 1974.
5. J. Millman and C.C. Halkias, Integrated Electronics, Chap. 15, McGraw–Hill Book Co., 1971.
6. P.R. Gray and R.G. Meyer, Analysis and Design of Analog Integrated Circuits, pp. 197–203, John Wiley and Sons, 1984.
7. J.E. Solomon, The Monolithic Op Amp: A Tutorial Study, IEEE J. Solid State Ccts, SC-9, pp. 314–332, Dec. 1974.
8. S.M. Sze, Physics of Semiconducutor Devices, pp. 320–340, John Wiley & Sons, 1969.
9. J.L. Moll, M. Tannenbaum, J.M. Goldey and N. Holonyak, p–n–p–n Switches, Proc. IRE, 44, pp. 1174–1182, 1956.
10. A. Blicher, Thyristor Physics, Chap. 1, Springer–Verlag, 1976.

Problems

1. (a) Design an opamp macromodel having the following specifications.
   CMRR = 100 dB
   Differential midband gain = 10^5
   Common mode input impedance = ∞
   Slew rate (both positive and negative) = 1 V/μ sec.
   Unity gain frequency = 1 MHz
   Output impedance = 50 Ω at all frequencies
   Input bias current = 250 nA
   Input current and voltage offset = 0
   (b) For this macromodel find the maximum differential input before the output saturates.

2. Consider the opamp macromodel with $C_E = 0$. It is connected in the voltage follower mode and a positive step of 5 V is given to the + input after it has been at 0 V for a long time. Show by writing node equations that

$$\frac{dv}{dt} = 2I_C/C_2$$

   assuming $G_a R_C = 1$. Would this expression for dv/dt be any different if the + input was at some non-zero voltage before the step is applied ?

3. In order to find the output impedance for the opamp macromodel, the Thevenin equivalent was found. Derive the expressions for V_{OC} and I_{SC} given in the text.

4. The input $v_+(t)$ for an opamp connected in the voltage follower mode is shown. Find $v_o(t)$ using the macromodel. Assume macromodel parameter values given in Table 7.1.

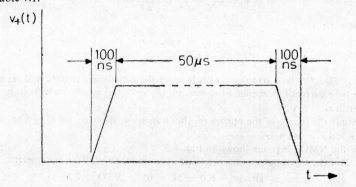

Fig. P. 7.4

5. Verify that the opamp specifications and macromodel parameters match in Table 7.1.

6. Let the opamp of problem 1 have a phase margin of 60°. Calculate C_1 to meet this requirement. What effect does C_1 have on the open loop differential gain at 1 KHz?

7. Assume opamp macromodel parameters given in Table 7.1 but omit the voltage and current limiting features. What are the maximum possible values for the output voltage and the output current from the macromodel?

Programming Assignments

Do the following assignments using a circuit simulation program like SPICE. Choose appropriate parameters for device models.

1. Plot $v(t)$ for the circuit shown when $v_s(t)$ switches abruptly from 5 V to -5 V. (Note that a finite rise time and fall time must be specified in SPICE and most other programs). The diode parameters are $I_S = 10^{-11}$ mA, $\tau_P = 5 n$ sec $(N_A \gg N_D)$, $C_{JO} = 50$ pF, $\phi_0 = 0.9$ V and $m = 0.5$.

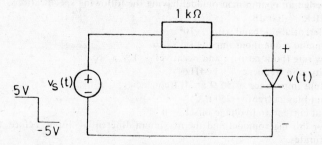

Fig. PA. 7.1

2. (a) Calculate the delay time, fall time, storage time and rise time for the transistor inverter shown in Fig. 7.3(a) with the following transistor parameters.

$BF = 100$ $BR = 1$ $IS = 2.0 \, E-16$ $TF = 0.1$ NS $TR = 10$ NS

$CJE = 0.1$ PF $VJE = 0.9$ V $MJE = 0.5$ $CJC = 0.05$ PF

$VJC = 0.7$ V $MJC = 0.5$

(b) The storage time can be almost completely eliminated by putting a capacitor C across the 10 K base resistor of Fig. 7.3(a). The value of the capacitor is given by

$$C = C_{SP} = \tau_S \frac{(I_b - I_{beos})}{\Delta V}$$

where

$$\tau_S = \frac{\alpha_F(\tau_F + \alpha_R \tau_R)}{1 - \alpha_F \alpha_R}$$

$(I_b - I_{beos}) =$ base overdrive with I_b being the actual base current and I_{beos} being the base current at the edge of saturation. $V_1 =$ initial steady state voltage across C with $v_s = 5$ V.

Study the nature of the output of the transistor inverter for $C = C_{SP}$, $C = 10$ C_{SP} and $C = 0.1$ C_{SP}.

3. For the NMOS inverter shown in Fig. 7.5

(a) Plot the input-output characteristic assuming the following parameters.

$$\text{Driver} - KP = 5E - 05 \qquad VTO = 2 \text{ V}$$
$$\text{Load} - KP = 1E - 05 \qquad VTO = -2 \text{ V}$$

(b) If you want to include the source to substrate bias effect for the depletion load, what parameter would you add? Repeat (a) by taking a typical value for this parameters.

(c) If a rise time of less than 1 n sec is required for a 0.02 pF load what should KP of the load transistor be changed to keeping other parameters the same ? What effect does this have on the input-output characteristic?

4. Repeat programming assignment 3 for the alternative type of NMOS inverter shown.

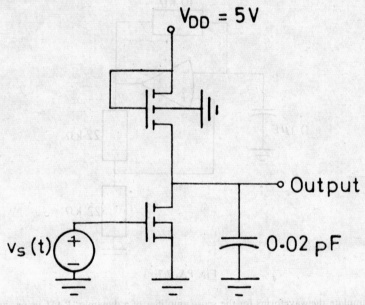

Fig. PA. 7.4

5. For an opamp connected in the voltage follower mode, find the pulse response and verify that it is similar to Fig. 7.13(a) Use macromodel parameters from Table 7.1.

6. Design a common emitter amplifier, as shown in Fig. PA 7.6, with the transistor

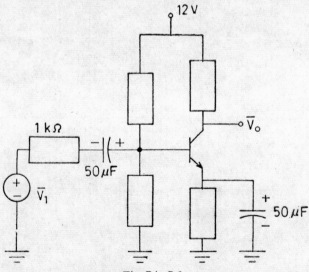

Fig. PA. 7.6

biased at $I_c = 2$ mA, and $V_{CE} = 4$ V. Assume transistor parameters from programming assignment 2. Midband voltage gain required is 100. Plot the frequency response and find the 3 dB points. Verify approximately by hand calculation. What important feature of the BJT small signal model is left out in the transistor parameters chosen?

7. Simulate the opamp astable circuit shown using opamp macromodel parameters from Table 7.1.

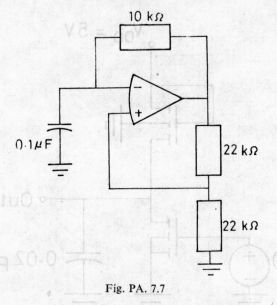

Fig. PA. 7.7

8. Simulate the waveforms for the sense amplifier of a dynamic *RAM* using just one transistor per memory cell. Use the circuits given in the following paper.
K.V. Stern et al, Storage Array and Sense/Refresh Circuit for Single Transistor Memory Cells, IEEE J. Solid State Ccts, SC-7, 5, pp 336–340, 1972.

8

Logic Simulation

So far we have dealt mainly with circuit analysis. Full fledged circuit analysis, though rigorous and complete, is impractical for very large circuits. The transient analysis of any circuit containing more than a hundred transistors is very time consuming. Analysing a digital circuit containing thousands of gates is out of the question. Fortunately, a complete circuit analysis of such a system is often unnecessary. In the early stages of design one is merely checking the logic. A non-numeric simulation at the logic level dealing with just '1s' and '0s' is a lot easier and meaningful. Not only are the computations non-numeric but the logic elements like gates are independent units. The output of a gate is strictly determined by its input and not what its output is connected to.

This kind of simulation can be done at many levels. One can have a behavioural description of the digital system in a high level language similar to PASCAL. Alternatively, one can describe a system in terms of counters, flip-flops, etc. This is done in RTL languages like AHPL ([1], [2]). A gate level simulation is at a lower level. This chapter deals mainly with gate level simulation. Often these levels overlap. The simulator LAMP [3], for example, enables one to hierarchically build up from the gate level. The AHPL simulator discussed in Section 8.9 allows parts of the system to be described in PASCAL.

Apart from verifying the logic, logic simulation is extensively used in fault simulation. The designer of a large digital system must provide test inputs which will help to test and locate as many of the faults as possible. Extensive logic simulation for various faults is necessary to generate these inputs. Many programs exist for automatic test generation [4].

Logic simulators, broadly speaking, are of two kinds—event driven and compiler driven. Gate level simulators are usually event driven. Here, in response to a change in any input the changes in the circuit are traced. Only the affected gates i.e. those in the signal flow path are analysed. In a compiler driven simulator, the input description is a language. This is compiled and executed as a program in a high level language. The gate level simulator described in Section 8.7 is event driven and is preferred at the gate level because dormant parts of the circuit are not analysed. Fault simulation is also done with gate level simulators.

Any event driven logic simulator would consist of the following sections:

(a) Parser to analyse the input description.
(b) Data base generator which creates a number of linked lists from the input description. Sub-circuits would be expanded. Once this data base of the digital system is created, various kinds of simulation runs can be performed.
(c) Actual simulator which uses the data base to analyse the system in response to any event or change in input.
(d) Output unit to print/plot various results.

8.1 Example of an Input Description Language

The logic simulator LAMP [3] uses a language called LSI-LOCAL to describe the circuit to be analysed. An exclusive-*OR* circuit and a full adder are described here using a similar language. It is given here to indicate how the input description may be given. Figure 8.1 shows the exclusive-*OR* circuit. The various gates have been given names A', B', AB', BA' and AXB. The output of a gate may also be referred to by the name of the gate itself. Alternatively, the output may be given a separate name, for example, the ouptut of gate AXB is called X. The input description is as follows:

$$\begin{array}{ll}
\text{CKTNAME} : & XOR; \\
\text{INPUTS} : & A, B \\
\text{OUTPUTS} : & X; \\
\text{NOT} : & A', A; \\
& B', B; \\
\text{NAND} : & AB', (A, B'); \\
& BA', (B, A'); \\
& AXB, (AB', BA'), X;
\end{array}$$

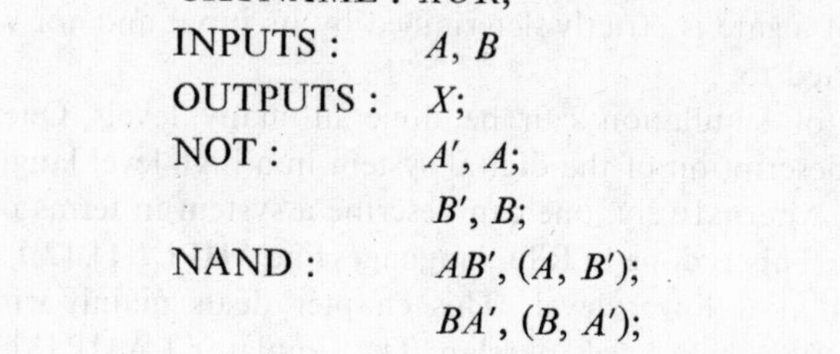

Fig. 8.1 Exclusive-*OR* circuit with gates named

In the description of the gates, the first term is the gate name, the second the set of inputs and the third the output. The output name is optional. Where there is more than one input, the set of inputs is enclosed by parentheses.

The description can be hierarchically built up. We can define the exclusive-*OR* as a block and use it to describe a full adder. A full adder with inputs

A, B and C (carry in) has outputs S (sum) and CO (carry out) defined as

$$S = A\bar{B}\bar{C} + \bar{A}B\bar{C} + \bar{A}\bar{B}C + ABC$$

$$= \bar{C}(A \oplus B) + C(\overline{A \oplus B}) = C \oplus (A \oplus B)$$

$$CO = AB + CA + CB$$

Here the symbol $\oplus$ stands for the exclusive-OR operation. The symbol for the exclusive-OR block is given in Fig. 8.2(a). Figure 8.2(b) shows the circuit for a full adder which uses the exclusive-OR block. The input description is as given below:

CKTNAME:	ADDER1
INPUTS:	A, B, C
OUTPUTS:	S, CO
XOR:	AXB, (A, B), X;
	D, (X, C), S
NOT:	A', A;
	B', B;
NAND:	ANB, (A, B);
	$AORB$, (A', B');
	$AORBNC$, $(AORB, C)$;
	CO, $(ANB, AORBNC)$;

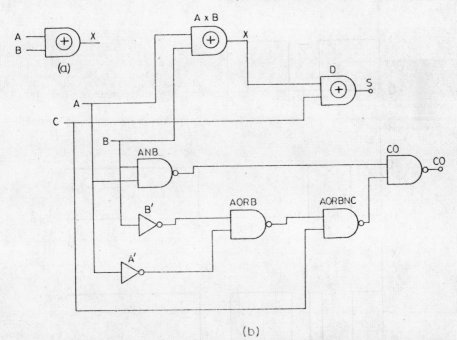

(a)

(b)

Fig. 8.2   (a) Exclusive-OR defined as a block. (b) Exclusive-OR block used in the full adder.

The gates have been named using the fact that gates at even levels (from the output) behave as AND gates. Gates at odd levels behave as *OR* gates with any fresh input at such a level appearing complemented. This is strictly true for a system only having NAND gates. Here we can assume that inverters have been realised with NAND gates.

While input descriptions of the kind given above have all the interconnection information, this information is not in a convenient form. For example, the fanout of a gate is not given in the gate specification. One has to search the inputs of all gates to find the fanout. It is also often necessary to find out other gates where a particular input of a given gate goes to. From a limited (though convenient to user) interconnection description a data base or a set of linked lists are first constructed. Once these lists are constructed, event driven simulation proceeds much faster.

8.2 Initial State Analysis

Logic simulation usually means finding the transient response to a set of time varying inputs. In the simplest case, these inputs switch back and forth between 0 and 1 at specified times. Output variations, again between 0 and 1, are to be found as a function of time. For the exclusive-*OR* circuit, input waveforms for *A* and *B* and the output *X* could be as shown in Fig. 8.3. It

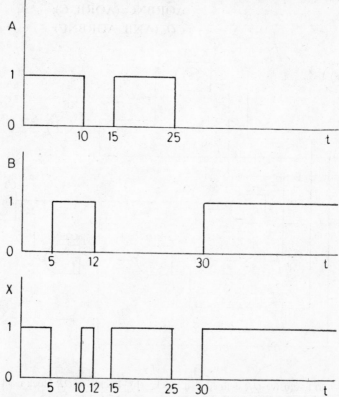

Fig. 8.3   Sample input and output waveforms for the exclusive-*OR* circuit.

was pointed out in Chapter 5 that transient analysis must be preceded by a DC analysis at $t = 0$. In other words, the initial state must be known before one can do a transient analysis. The initial state is also, in any case, part of the response. Only, it can be found more easily as initial conditions like capacitor voltages and inductor currents are specified. As in circuit analysis, logic simulation is usually preceded by an initial state analysis. Outputs of all elements are found at $t = 0$. Memory elements, like NAND or NOR latches, must have this initial state specified. Of course, the various inputs must be specified at $t = 0$ and at all further times.

Consider the exclusive-OR circuit shown in Fig. 8.4. Let us assume the initial inputs to be $A = 1$, $B = 0$. It is necessary to find the outputs of all gates at $t = 0$. This is done in many iterations. In the first iteration no gate output is known. All unknown outputs are indicated by 2. In each iteration, the output of a gate is found by taking input values from the previous iteration. Initial state analysis is completed when all outputs are found and when two successive iterations give the same values. Results for the exclusive-OR circuit are given in Table 8.1. Iterations 4 and 5 give the same output values and we stop after iteration 5. Now, if inputs A and B vary as function of time, the outputs can be calculated as described in the next section.

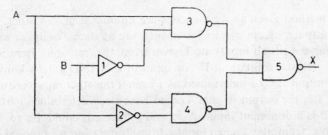

Fig. 8.4 Exclusive-OR circuit redrawn with index number for gates.

Table 8.1 Initial state analyis of exclusive-OR circuit (Fig. 8.4)

Primary Input/ Gate output	Iteration 1	Iteration 2	Iteration 3	Iteration 4	Iteration 5
A	1	1	1	1	1
B	0	0	0	0	0
1	2	1	1	1	1
2	2	0	0	0	0
3	2	2	0	0	0
4	2	2	1	1	1
5	2	2	2	1	1

As a second example, consider the J-K flip-flop shown in Fig. 8.5. It is not of the master-slave kind and its output may oscillate if $J = K = Clk = 1$. We will not worry about this and will attempt to find the initial state corresponding to $J = K = 0$ and $Clk = 1$. As it is a sequential circuit the values of the NAND latch; Q and $\bar{Q}$, need to be specified. Let $Q = 1$, $\bar{Q} = 0$. Then the analysis proceeds as in Table 8.2.

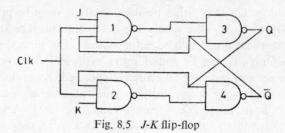

Fig. 8.5 *J-K* flip-flop

Table 8.2 Initial state analysis of J-K flip-flop (Fig. 8.5)

Primary input/ gate output	Iteration 1	Iteration 2	Iteration 3
J	0	0	0
K	0	0	0
Clk	1	1	1
1	2	1	1
2	2	1	1
3	1	1	1
4	0	0	0

In the method given above the gates are numbered in any order. Initially all gate outputs, except memory elements, are at state 2 or unknown. A gate is not analysed till all inputs are known from the previous iteration. This could be relaxed somewhat. If one input of a NAND gate is known to be zero, its output can be determined as 1 even if the other inputs are unknown. In Table 8.1, the output of gate 4 could have been determined in iteration 2 itself as *B* is a dominant input. A second refinement could be to trace the signal flow from the known inputs. In each iteration we proceed one level and do not attempt to determine all gate outputs. In Fig. 8.4 the inputs lead to gates 1, 2, 3 and 4. In iteration 2, we would only attempt to analyse these gates. Once these are determined we analyse the gates, these in turn lead to. This requires more programming effort but is faster. Moreover, the information to trace the signal flow is anyway needed for the simulation as a function of time, This may be used. Even so, initial state analysis is done just once and a slower, but simpler, technique may be preferred.

8.3 Unit Delay Simulation

In this section we assume that all gates have the same delay. This delay is taken as the unit delay, whatever be its value. The time step of analysis is chosen to be equal to this unit delay. It is rather simplistic to represent various delays in a gate by one delay. One can talk of rise time, fall time, low to high propagation delay, high to low propagation delay, etc. A logic simulation can give only limited timing information, and it is alright to talk of just one delay in a simple unit delay simulator. We will again take the exclusive-OR circuit shown in Fig. 8.4 as an example. At the end of the input description the variation of the inputs as a function of time is given. This information may be fed as follows:

$$A = 1, B = 0$$

$$A \quad 10 \quad 0 \quad 15 \quad 1 \quad 25 \quad 0$$

$$B \quad 5 \quad 1 \quad 12 \quad 0 \quad 30 \quad 1$$

$$\text{PRINT } X \quad 0 \quad 40$$

The first line gives the initial state. The second line indicates how input A varies as a function of time. At time step 10, A changes to 0, at time step 15 to 1 and at time step 25 to 0. Similar information for input B is contained in the third line. Finally, the time period over which the analysis is to be done and the output to be printed are given. What is given here is one simple way of feeding in the variation of inputs and some simulation commands. It is not necessarily the best or only way.

We can now analyse the circuit as a function of time in a manner similar to initial state analysis. At each time step we use input values from the previous time step to calculate the output. This is to take care of the unit delay. If the response to any input has settled down we can skip ahead to the next time step where some input changes. Table 8.3 shows the analysis for the exclusive-OR of Fig. 8.4. The variation of inputs is as given above. The initial state analysis is not shown. After the initial state analysis we can directly skip to time step 5 as no input changes in between and the circuit is in steady state. At time step 5 input B changes. We get the same outputs for time steps 8 and 9, again indicating the circuit has stabilised. But the next input change is at time step 10. So no time steps can be skipped.

Table 8.3 Unit delay simulation for exclusive-OR

Primary input/gate output	Time step 0	Time step 5	Time step 6	Time step 7	Time step 8	Time step 9
A	1	1	1	1	1	1
B	0	1	1	1	1	1
1	1	1	0	0	0	0
2	0	0	0	0	0	0
3	0	0	0	1	1	1
4	1	1	1	1	1	1
5	1	1	1	1	0	0

While the above scheme is workable, it is very inefficient. It is especially so for very large circuits. The reason is that we analyse all gates at all time steps. Usually, an input change affects only a small part of a large circuit. It should be necessary to analyse only the affected gates. Even in the simple exclusive-OR circuit, an input change in B does not affect gate 2 in any way. Also, an input change in B takes time to propagate to gate 5. It is not necessary to analyse gate 5 until one of its input has changed. All this means that we must trace the signal flow path. A selective trace is to be done to analyse only those gates lying on the signal flow path. Information regarding the interconnections of the gates must be stored in a suitable fashion to do this conveniently. Actual data structures to do this will be considered in Section 8.5. Here we will assume that the interconnection details are readily available.

A simple analysis scheme based on selective trace is given in Reference [3], [5]. Two lists A and B are maintained. At the beginning of any time step t_n, list A contains the gates to be analysed at t_n. These gates are then analysed. The fanout of every gate whose output changes is entered into list B. List B at the end of t_n, contains the gates to be analysed in t_{n+1}. Lists A and B are exchanged and analysis proceeds to the next time step. If list B is empty at the end of some simulation step, we stop simulation unless there are input changes. Applying this technique to the exclusive-OR circuit of Fig. 8.4, we get the following contents for lists A and B (Note: Lists A and B should not be confused with inputs A and B).

Beginning of t_5: list A has input B
    End of t_5: list B has gates 1 and 4
Beginning of t_6: list A has 1 and 4
    End of t_6: list B has 3
Beginning of t_7: list A has 3
    End of t_7: list B has 5
Beginning of t_8: list A has 5
    End of t_8: list B is empty

When list B becomes empty we stop simulation till some input changes.

Table 8.4 is similar to Table 8.3 except that it shows only those gates which are analysed using selective trace as described above, Naturally, the number of gates analysed is far less even for the simple 4 gate exclusive-OR circuit.

Table 8.4   Analysis of Exclusive-OR using Selective-trace

Primary input-gate output	Time step 0	Time step 5	Time step 6	Time step 7	Time step 8
A					
B		1			
1			0		
2					
3				1	
4			1		
5					0

Figure 8.6 gives the flow chart for the unit delay simulator using selective trace. It is assumed that no new input will change till the circuit has stabilised after previous input changes. Therefore, one checks for input changes only after list B becomes empty. If the circuit does not stabilise after a given

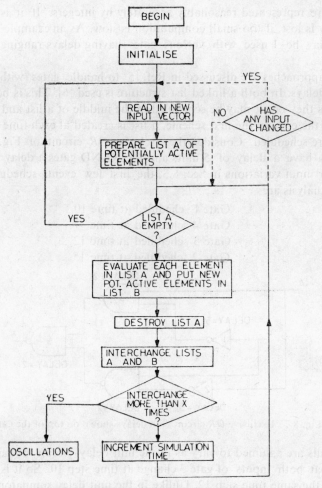

Fig. 8.6 Flowchart for a unit delay simulator

number of time steps, it is assumed that there are oscillations. Oscillations are detected if lists A and B are exchanged more than a given number of times. Most digital circuits would be so designed that new input changes would be disallowed till the old changes have percolated through the various levels to the output. If input changes are to be allowed sooner, then the flow would be modified as indicated by the dotted lines in Fig. 8.6. However, it would no longer be meaningful to check for oscillations.

8.4 Simulation with Gates Having Arbitrary Delays

It is unrealistic to expect that all gates would have the same delay. There are different kinds of gates with different kinds of loadings. Gate delays in a circuit may easily vary by an order of magnitude. The simulation procedure is then quite different. First the unit of delay must be so chosen that all gate delays are represented reasonably accurately by integers. If it is too large accuracy is lost, if too small computation is slow. As an example the unit of delay may be 1 nsec with various gates having delays ranging from 2 to 20 units.

Two approaches are discussed in Ref. [5] to handle gates with arbitrary integer delays. In both a linked list structure is used [6]. This is because new events as they occur may be scheduled in the middle of a list and not necessarily at the end. In the first scheme, a list is created at each time step where events are scheduled. Consider the exclusive-*OR* circuit of Fig. 8.7. The inverters have a delay of 5 units and the NAND gates a delay of 2 units. Then for input variations in Sec. 8.3, the first few events scheduled as we do the analysis are

Gate 1 scheduled at time 10
Gate 4 scheduled at time 7
Gate 3 scheduled at time 12
Gate 2 scheduled at time 15

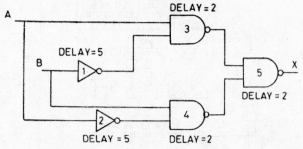

Fig. 8.7   Exclusive-*OR* circuit with delays shown on top of the gates

The inputs are assumed to vary as in the unit delay circuit of last section. Note that both inputs of gate 3 change at time step 10. So it is scheduled twice at the same time step 12. Unlike in the unit delay simulator, the output of gate 3 never goes to 1 because the input *A* changes to 0 by the time the other input from gate 1 changes to 1. The array of various time steps would be given by *T* below and the pointers which give the sequence of time steps would be given by *TP*.

$$TP \text{ START} = 2$$

$$
\begin{array}{ll}
T(1) = 10 & TP(1) = 3 \\
T(2) = 7 & TP(2) = 1 \\
T(3) = 12 & TP(3) = 4 \\
T(4) = 15 & TP(4) = *
\end{array}
$$

The linked list described by *T* and *TP* is shown in Fig. 8.8. For each index *i* of array *T* we have a list of events scheduled at *T*(*i*). These events can be

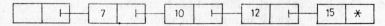

Fig. 8.8   Linked list showing times at which changes occur

in an ordinary list or linked list. A new event at $T(i)$ can just be added at the end of the list of the events scheduled at $T(i)$. The pointer array TP makes it easy to insert a new event scheduled at a new time step. Assume that an event is to be scheduled at time step 14. Then we just make $T(5) = 14$, $TP(3) = 5$ and $TP(5) = 4$. The main disadvantage with this scheme is that one must go through TP and T to insert a new event at a new time step or add it to the existing list at a old time. This may be time consuming.

In the second scheme we have a linked list of events originating from every time step. Many of these lists may be empty. The advantage is that if a new event at a new time step is to be scheduled we know how many time steps to advance from the current time step. Assume that in some circuit we have the following events.

Gates 5, 20, 10, 12, 25, 13 at time 12

None at time 13

None at time 14

Gates 21, 30, 5, 14, 12, 16 at time 15

At times 12 to 15 we have the array of gates G and array of pointers GP as

$$H(12) = 1$$

$G(12, 1) = 5$	$GP(12, 1) = 3$
$G(12, 2) = 20$	$GP(12, 2) = 5$
$G(12, 3) = 10$	$GP(12, 3) = 4$
$G(12, 4) = 12$	$GP(12, 4) = 6$
$G(12, 5) = 25$	$GP(12, 5) = *$
$G(12, 6) = 13$	$GP(12, 6) = 2$

$$H(13) = *$$
$$H(14) = *$$
$$H(15) = 3$$

$G(15, 1) = 21$	$GP(15, 1) = 2$
$G(15, 2) = 30$	$GP(15, 2) = *$
$G(15, 3) = 5$	$GP(15, 3) = 5$
$G(15, 4) = 14$	$GP(15, 4) = 6$
$G(15, 5) = 12$	$GP(15, 5) = 4$
$G(15, 6) = 16$	$GP(15, 6) = 1$

The headers from each time i should point to the leading element of the list

$$H(12) = 1$$
$$H(13) = *$$
$$H(14) = *$$
$$H(15) = 3$$

It has been assumed that the actual order of analysis is the order in which gates have been numbered. Actually gates scheduled at the same time can be analysed in any order. Now assume that when doing the simulation at time 9 an event is scheduled 5 time steps away. If the gate is say 28 then all one has to do is assign

$$G(14, 1) = 28, \qquad H(14) = 1, GP(14, 1) = *$$

The major disadvantage of this scheme is that we have empty lists at a large number of time steps. As we have seen in the case of the exclusive-OR, only a small percentage of the times may have events scheduled. Some storage space can be saved by using circular lists. If the largest delay is 99 units, we may consider just 100 time steps at a time. Let these correspond to 1 to 100. Assume we are at the 81st time step in simulation and an event is scheduled at time step 121. We can create a list now at time step 21 corresponding to time step 121. We no longer need lists for events scheduled before the 81st time step.

For the unit delay simulator, it was pointed out that input values from the previous time step must be used to find the outputs of a gate. The same principle applies here except that the inputs may be from a time several time steps earlier. Values from the current time step should not be used. Otherwise one cannot analyse circuits with feedback. If there were a NAND or NOR latch, one would be going round and round.

8.5 Data Structures for Logic Simulation

As input description of the kind given in Section 8.1 does not contain enough information to do a selective trace in response to an input event. The information is there, but not in a form readily usable. More information is usually extracted and stored in tables. Two data structures for storing the various attributes of a gate/element are discussed here. Both are based on the discussion in Ref. [5].

In the first, each gate is described by the following quantities:

1. Name
2. Index i
3. Output value
4. NFO—Number of fanouts
5. NFI—Number of fanins
6. FOL—Pointer to a list called IOLST. FOL points to first fanout. Next (NFO(i) — 1) locations of IOLST contain remaining fanouts of i.
7. FIL—Pointer to IOLST. Similar to FOL but for inputs.
8. Type (NAND, NOR, etc.)
9. Delay

IOLST is a list of the various outputs and inputs of all the gates or elements. It starts with the fanins of the first gate, then has the fanouts of the first gate, then the fanouts of the 2nd gate and so on. For the ith gate FOL(i)

points to the leading fanout element and FIL(i) to the leading fanin element. The remaining fanouts/fanins follow in sequence. Consider the master-slave J-K flip-flop with the gates numbered as shown in Fig. 8.9. FIL, FOL and IOLST for this circuit are as given below:

i	FIL(i)	FOL(i)	j	IOLST(j)
1	1	4	1	J
2	5	8	2	Clk
3	9	11	3	8
4	13	15	4	3
5	17	19	5	K
6	20	22	6	Clk
7	23	25	7	7
8	27	29	8	4
9	31	32	9	1
			10	4
			11	4
			12	5
			.	.
			.	.
			.	.
			31	Clk
			32	5
			33	6

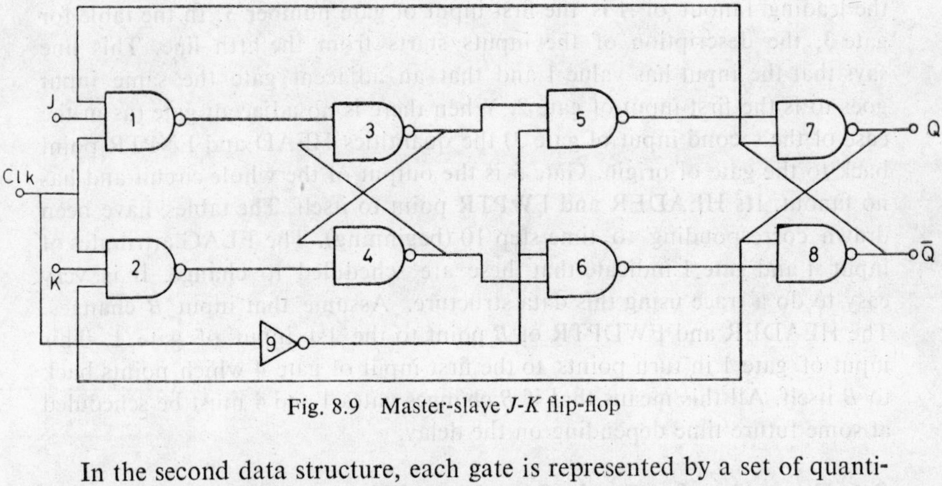

Fig. 8.9   Master-slave J-K flip-flop

In the second data structure, each gate is represented by a set of quantities as shown in Fig. 8.10.

The header and forward pointer (FWDPTR) point to the leading fanout element of the gate. The index number, name, output logic value, number of fanins, kind and delay are standard attributes. The flag indicates if the element has already been scheduled or not. A major convenience in this structure is that details of the inputs are provided in the table of the element

HEADER		FWDPTR
INDEX		NAME
OUTPUT LOGIC VALUE		NUMBER OF FANINS
FLAG	KIND	DELAY
VAL(i)	HEAD(i)	FNPTR(i)
i = 1		
i = 2		

Fig. 8.10 Table describing attributes of a gate

itself. The value of each input as well as the adjacent element the same input goes to, are given. Adjacent elements are pointed to by the attributes HEAD and FNPTR (fan pointer). For further details the reader is referred to Ref. [7].

This structure is best illustrated by an example. We will once again take the exclusive-OR of Fig. 8.7. The tables for the various gates/elements are as shown in Fig. 8.11. The indices for the gates are the gate numbers. Primary input A has index 6 and B index 7. The input tables of A and B are empty as these are primary inputs themselves and have no fanin. The first line of the table for A has HEADER = 1 and FWDPTR = 3. This means that the leading fanout of A is the first input of gate number 3. In the table for gate 3, the description of the inputs starts from the fifth line. This line says that the input has value 1 and that an adjacent gate the same input goes to is the first input of gate 2. When there is no adjacent gate (as in the case of the second input of gate 3) the quantities HEAD and FNPTR point back to the gate of origin. Gate 5 is the output of the whole circuit and has no fanout. Its HEADER and FWPTR point to itself. The tables have been drawn corresponding to time step 10 (beginning). The FLAG attributes of input A and gate 1 indicate that these are scheduled to change. It is very easy to do a trace using this data structure. Assume that input B changes. The HEADER and FWDPTR of B point to the 1st input of gate 1. This input of gate 1 in turn points to the first input of gate 4 which points back to B itself. All this means that if B changes gates 1 and 4 must be scheduled at some future time depending on the delay.

8.6 Example of a Logic Simulator [7]

A logic simulator called SIMLOG developed at IIT Kanpur is described here as an example. It takes any value of transport delay for an element and the basic scheme is as shown in Fig. 8.12. It is simpler than either of the schemes in Section 8.4 though a little less efficient. Any time an input to an

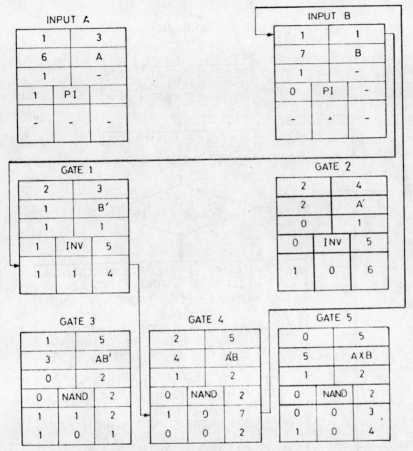

Fig. 8.11 Tables for the exclusive-*OR* circuit of Fig. 8.7 based on data structure of Fig. 8.10

element changes, the element number and delay are stored in a stack T. At each time step, elements in the stack are searched to see if any of them have a delay Δ equal to 1. Such elements are stored in list L_B and their outputs changed. Fanouts corresponding to these outputs are stored in list L_{t+1} so that at the next time step these elements can be entered into the stack T. The delay Δ for elements with $\Delta > 1$ is decremented by 1 at each time step.

The element descriptor record for each element includes the index number, name, number of fanouts NFO, number of fanins NFI, fanout list FOL, fanin list FIL, type of element and the delay. A global linked list is formed for all elements. Then each descriptor table is pointed to by an array of pointers. This helps in assessing and manipulating any data of any element.

The simulator is a 4-level simulator. In addition to logic levels 0 and 1, it has the unknown state represented by 2 and tristate or high impedance by 3. Pass transistors or transmission gates are represented as voltage controlled

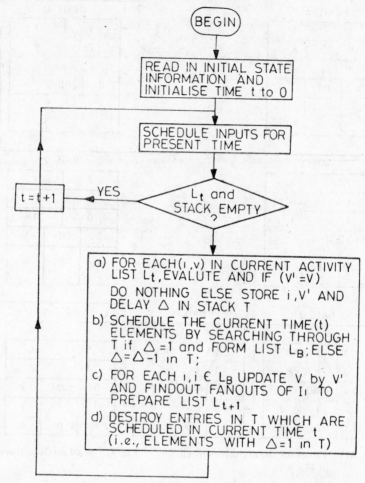

Fig. 8.12 Flow chart for simulator having different delays for different gates

switches. When the control input is 1, the input is available at the output; when the control is 0, the output is 3 or tristate. The truth table is as given below:

	CONTROLS			
INPUT	0	1	2	3
0	3	0	2	2
1	3	1	2	2
2	3	2	2	2
3	3	3	2	2

The output of two or more pass transistors or tristate elements can be connected to a bus. An element called BUS is also represented. It follows the following logic.

INPUT *A*

INPUT *B*	0	1	2	3
0	0	0	0	0
1	0	1	2	1
2	0	2	2	2
3	0	1	2	*

The * in the table means that the previous state is retained. Where a bus conflict occurs (i.e. one input is 1 and another 0), the 0 is assumed to dominate and the printed output indicates that a bus conflict occurred.

The above representation of pass transistors assumes that they are unidirectional. There are many applications where their bidirectional nature is important [10]. An example is the RAM cell shown in Fig. 8.13(b) C_0 and C_1 correspond to WRITE and READ signals. When READ alone is active, transmission is right to left onto the data bus from the cell. When C_0 (or WRITE) and C_1 are active transmission is left to right. Thus the pass transistor controlled by C_1 is bidirectional while that controlled by C_0 is unidirectional. Bidirectional pass transistors are taken care of by defining a

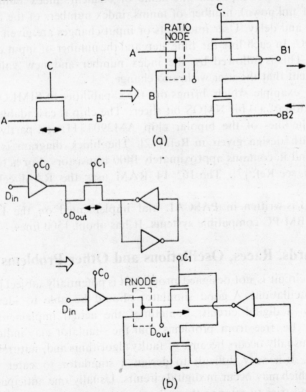

Fig. 8.13 (a) Representation of a bidirectional pass transistor in terms of two unidirectional ones and a special element called RNODE (b) Use of bidirectional model of (a) to describe a RAM cell

new element called RNODE [10]. The RAMCELL is modified as shown in Fig. 8.13(b) using RNODE, thereby converting bidirectional pass transistors to unidirectional ones. The logic for RNODE is as follows assuming INPUT A to be stronger than INPUT B (i.e. it has more driving capability).

	INPUT A			
INPUT B	0	1	2	3
0	0	1	2	0
1	0	1	2	1
2	0	1	2	2
3	0	1	2	3

Another feature of SIMLOG is that it allows sub-circuits. If some block is repeatedly used (say a flip-flop or a memory cell) it can be described once and called as often as necessary with different external links. Only one level of sub-circuits is allowed in SIMLOG.

The full adder circuit shown in Fig. 8.14(a) was analysed and is given as an example. Figure 8.14(b) shows the input listing and Fig. 8.14(c) the output. Each line of input has the name of element, index number, initial value (2 if unknown), number of fanins, index numbers of the fanins, type of element and delay. User interrupts or input changes are given in the end after ENDC. In each line the time step and the number of input changes are given first. This is followed by the index number and new value of each input element that the user wishes to change.

Another example which brings out the capabilities of SIMLOG better is the simulation of a 4 bit NMOS bit slicer. The chip is cascadable, designed partly on the lines of the bipolar chip AM2903 [11] and partly on that of the data path design given in Ref. [12]. The block diagram is shown in Fig. 8.15 and it contains approximately 1400 transistors. For actual simulation details, see Ref. [7]. The 16×14 RAM uses the RAM cell given in Fig. 8.13.

SIMLOG is written in PASCAL and implemented on the DEC 10/90 as well as IBM PC compatible systems. It has about 1500 lines.

8.7 Hazards, Races, Oscillations and Other Problems

If a digital circuit is not designed properly it is potentially subject to hazards, races and oscillations. A good simulator should be able to identify these problems in a digital circuit. Sometimes the actual implemented digital circuit may be free from problems; but the simulator may indicate a problem. This usually occurs because of faulty algorithms and, naturally, should be avoided. It is very difficult to design a simulator to cater to all the problems which may occur in digital circuits. Usually one anticipates certain problems and tries to cater to these.

A circuit where a hazard could occur is shown in Fig. 8.16(a). Gate B is a non-inverting buffer and gate C is an inverter. Let A go from 0 to 1 and back to 0. If gate B is a little slower than C, Y goes momentarily to 0 after

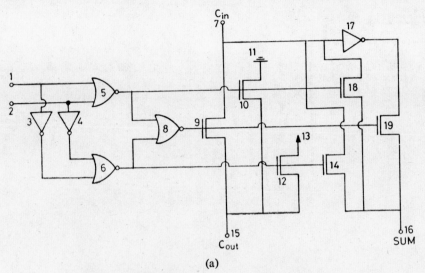

(a)

```
*  THIS DESCRIBES THE FULL ADDER CIRCUIT.
*  THERE IS NO SUBCIRCUIT DESCRIPTION.
*  STARTS WITH MAIN DESCRIPTION.
*  THERE ARE 19 ELEMENTS IN THIS CIRCUIT.
*  THE CIRCUIT IS SIMULATED FOR 120 TIME STEPS
INPT 1 1 0 BINP 0
INPT 2 0 0 BINP 0
INVT 3 1 1 2 NOT 1
INVT 4 0 1 1 NOT 1
NOR1 5 2 2 1 2 NOR 1
NOR2 6 2 2 3 4 NOR 1
CINB 7 1 0 BINP 0
NOR3 8 2 2 5 6 NOR 1
BUF1 9 2 2 7 8 BFFR 1
BUF2 10 2 2 11 5 BFFR 1
LOWI 11 0 0 BINP 0
BUF3 12 2 2 13 6 BFFR 1
HIGH 13 1 0 BINP 0
BUF4 14 2 2 7 6 BFFR 1
BUS1 15 2 3 9 10 12 BUS 0
BUS2 16 2 3 14 18 19 BUS 0
INVT 17 2 1 7 NOT 1
BUF5 18 2 2 7 5 BFFR 1
BUF6 19 2 2 17 8 BFFR 1
*  NOS. 1,2 & 7 ARE INPUTS.
*  SUM IS NO.16,CARRY IS NO.15.
ENDM
*  THERE WILL BE NO SUBCIRCUIT CALLS.
*  SO PLACE ENDC IN NEXT LINE.
*  THERE ARE USER INTERRUPTIONS IN THIS CASE.
*  INTERRUPTIONS ARE DESCRIBED AFTER ENDC.
ENDC
10 1 7 0
30 2 1 0 2 1
50 1 7 1
70 1 1 1
90 3 1 0 2 0 7 0
0
```

(b)

```
TY 5.2B
[23:12:47]
THE GIVEN TABLE IS :
```

INDEX	NAME	VALUE	NFO	NFI	FOL												FIL											KIND	DELAY
1	INPT	1	2	0	4	5	0	0	0	0	0	0	0	0	0	0	0	0	0	0	0	0	0	0	0	BINP	0
2	INPT	0	2	0	3	5	0	0	0	0	0	0	0	0	0	0	0	0	0	0	0	0	0	0	0	BINP	0
3	INVT	1	1	1	6	0	0	0	0	0	0	0	0	0	0	0	0	0	0	0	0	0	0	0	0	NOT	1
4	INVT	0	1	1	6	0	0	0	0	0	0	0	0	0	0	0	0	0	0	0	0	0	0	0	0	NOT	1
5	NOR1	2	3	2	8	10	18	0	0	0	0	0	0	0	1	0	0	0	0	0	0	0	0	0	0	NOT	1
6	NOR2	2	3	2	8	12	14	0	0	0	0	0	0	0	1	2	0	0	0	0	0	0	0	0	0	NOR	1
7	CINB	1	4	0	7	14	17	18	0	0	0	0	0	0	3	4	0	0	0	0	0	0	0	0	0	NOR	1
8	NOR3	2	2	2	9	19	0	0	0	0	0	0	0	0	5	6	0	0	0	0	0	0	0	0	0	BINP	1
9	BUF1	2	1	2	15	0	0	0	0	0	0	0	0	0	7	8	0	0	0	0	0	0	0	0	0	NOR	1
10	BUF2	2	1	2	15	0	0	0	0	0	0	0	0	0	11	5	0	0	0	0	0	0	0	0	0	BFFR	1
11	LOW1	0	1	0	10	0	0	0	0	0	0	0	0	0	0	0	0	0	0	0	0	0	0	0	0	BFFR	1
12	BUF3	2	1	2	15	0	0	0	0	0	0	0	0	0	13	4	0	0	0	0	0	0	0	0	0	BINP	0
13	HIGH	1	1	0	12	0	0	0	0	0	0	0	0	0	0	0	0	0	0	0	0	0	0	0	0	BFFR	0
14	BUF4	2	1	2	16	0	0	0	0	0	0	0	0	0	7	6	0	0	0	0	0	0	0	0	0	BINP	0
15	BUS1	2	0	3	0	0	0	0	0	0	0	0	0	0	9	10	12	0	0	0	0	0	0	0	0	BFFR	1
16	BUS2	2	0	3	0	0	0	0	0	0	0	0	0	0	9	10	12	0	0	0	0	0	0	0	0	BUS	0
17	INVT	2	1	1	19	0	0	0	0	0	0	0	0	0	14	18	19	0	0	0	0	0	0	0	0	BUS	0
18	BUF5	2	1	2	16	0	0	0	0	0	0	0	0	0	7	0	0	0	0	0	0	0	0	0	0	NOT	1
19	BUF6	2	1	2	16	0	0	0	0	0	0	0	0	0	17	8	0	0	0	0	0	0	0	0	0	BFFR	1

evaluation results

```
AT TIME?      GATE  16.        GATE  15.
********      ********        ********

     0          0                        1
    11          0             0 -------- 1
    12          -------- 1    0
    31          1             0
    51          1             ----■-- 1
    52          0 ------- 1    1
    71          0             1
    72          0             1
**    16
    73          0             1
    74          -------- 1    1
    91          0 ------- 1    1
**    15
    92          0             0 ----■-- 
    93          0             0
```

```
**   XX :INDICATES BUS CONFLICT AT GATE/ELEMENT NO. XX .
```

```
TOTAL TIME DELAY=  120                    THE EXECUTION TIME :    1.402 SECONDS
```

(c)

Fig. 8.14   The full adder circuit analysed by SIMLOG.

the 0 to 1 transition of A. If C is a little slower than B, Y goes to 0 momentarily after the 1 to 0 transition of A. Let both gates B and C have a declared delay of 1 unit when simulated. Then the potential hazards during the two transitions would be missed by a unit delay simulator. Of course if one gate is declared to have a larger delay, the hazard would not be missed. A scheme is suggested by Eichelberger [8] to detect static hazards. Here we assume that an input goes through an unknown state 4 during a transition from 0 to 1 or 1 to 0. Input A in Fig. 8.16(a) would now have a sequence 04140. We find the steady state output Y due to each of these input values in the sequence. We would get 14141. Any time a sequence 141 or 040 occurs, it indicates a hazard i.e. a momentary or transient pulse which may produce an error.

Races and oscillations are more of a problem in asynchronous sequential circuits. The standard form of a synchronous circuit with a two phase clock is as shown in Fig. 8.17. Either ϕ_1 or ϕ_2, but not both, is high at a given instant. When ϕ_1 is high and ϕ_2 is low, the output of the combinational circuit may have transient spikes. Only when the circuit stabilises does ϕ_2 go high latching the output of the combinational circuit. When ϕ_2 is high, the

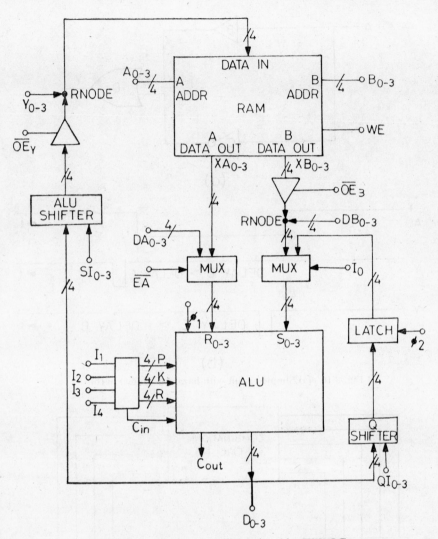

Fig. 8.15   The bit slice chip analysed by SIMLOG

inputs to the combinational circuit cannot change (as ϕ_1 is low) thereby
ensuring that the output is stable. Many problems of oscillations and races
are solved by this arrangement. A two phase clock of this kind is used in
most digital circuits.

Another standard problem posed is that of the NAND or NOR latch.
Consider the NAND latch shown in Fig. 8.18. Let both S and R be given
the input sequence 01. The final state then depends on which NAND gate is
faster. If the upper gate is faster we end up with $Q = 0$ and $\bar{Q} = 1$. If the
lower gate is faster we get $Q = 1$, $\bar{Q} = 0$. In a simulator both gates would
be given the same delay. Therefore, both gates would be scheduled together
and the output will oscillate, A race condition appears as an oscillation in

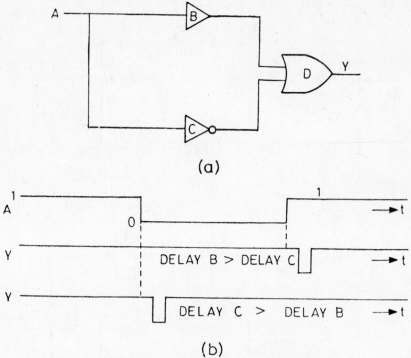

Fig. 8.16   (a) Simple circuit with hazard as shown in (b)

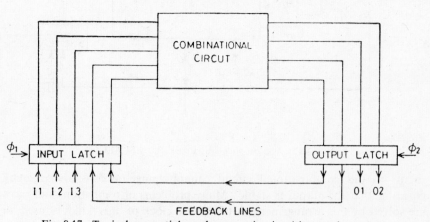

Fig. 8.17   Typical sequential synchronous circuit with two phase clock

the simulation. In the LAMP simulator [3] this race condition for a NAND latch is detected by the following conditions.

(a) The newly calculated, but not yet assigned, outputs of both gates are simultaneously 0.

(b) Both gate outputs are scheduled to be changed at the present time step.

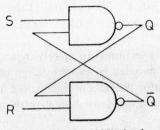

Fig. 8.18   A NAND latch

Reconvergent fanouts also lead to problems in simulation. The circuit of Fig. 8.16(a) is an example of reconvergent fanout. Two or more inputs of a gate originate from the same gate/input but through different paths. Both inputs to the OR gate D originate from A but through different paths. We will now introduce the concept of ambiguity delay. Instead of assigning a fixed delay to each gate we say that the delay can be between x_{min} and x_{max} units. The output of the gate is indeterminate between x_{min} and x_{max} units after an input change (if the output is to change for this input transition). In Fig. 8.16(a) let B have a delay between 5 and 10 units and C a delay between 12 and 17 units. Let input A itself be the output of some gate so that A makes a transition from 0 to 1 between 0 and 5 units. A simple analysis indicates that gate B changes from 0 to 1 somewhere between 5 and 15 units. Gate C changes from 1 to 0 between 12 and 22 units. These are shown in Fig. 8.19. This implies that between 12 and 15 units both B and C could be zero implying a hazard at output Y. But we know that gate

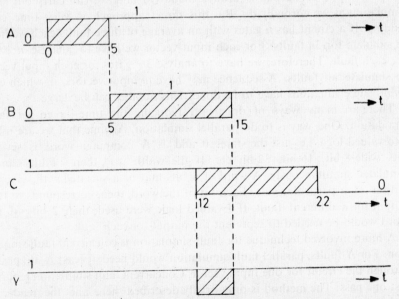

Fig. 8.19   Hazard predicted for circuit of Fig. 8.16 because of reconvergent fanout when no hazard exists

B is always faster than gate C. The output for these delays can never become zero. In short, the simulation may predict hazards for reconvergent fanouts when none exist.

The discussion in this section is to merely indicate pitfalls in simulation. Some of these can be avoided by proper design of the digital system. One may not care if a simulator predicts problems like hazards where none exist. Depending on the design, one may design overly pessimistic simulators or otherwise.

8.8 Fault Simulation

One of the major applications of logic simulation is in fault simulation. Fault simulation is useful in designing test sequences to test an IC which has been manufactured. There are various kinds of faults which one has to look for in testing. Different technologies like CMOS and TTL may have faults peculiar to them. Fortunately, most faults which occur in a digital circuit are equivalent to two classical faults. These are the 'stuck at 1' and the 'stuck at 0' faults which correspond to some input or output stuck at 1 or 0. Stuck open faults are common in MOS ICs. The conversion of these to equivalent stuck at 1/0 faults is discussed in reference [12]. A circuit may have more than one fault at a time. However, tests generated for single faults usually detect multiple faults as well. Most techniques for fault analysis are therefore restricted to single stuck at 1/0 faults. We shall briefly discuss fault simulation under these restrictions.

Simulating a digital circuit when it is stuck at 1/0 is easy. The corresponding input or output is always made 1/0 and we can carry out logic simulation as we normally do. But this direct approach is very time consuming. If a circuit has n gates with an average of three faults each we have to simulate for $3n$ faults. For each input vector we have to analyse n gates for each fault. Therefore, we have to analyse $3n^2$ gates for each input vector to simulate all faults. A sequence may have m input vectors, in which case $3n^2m$ gate evaluations have to be done. This number can be large.

There are many ways of drastically reducing the time taken for fault simulation. One way is to do parallel simulation. Assume that we are using two-valued logic i.e. just the states 0 and 1. A computer word is several bits wide-8-bit, 16-bit, 32-bit etc. If the width is w, then w faults can be simulated simultaneously. A word of w bits may be associated with, say, the output of a particular gate. Each bit of the word, then, corresponds to that output for a different fault. If 4-valued logic were used, then 2 bits of the word would be needed to represent the output for each fault.

A more involved technique for fault simulation is concurrent fault simulation. For N faults, parallel fault simulation would need at least N/W passes through the circuit for one input vector. Concurrent fault simulation requires just one pass. The method is only briefly described here and the reader is directed to reference [5] for more details. In concurrent fault simulation, a super fault list is created for each element or gate corresponding to an input vector. As an example, consider the exclusive-*OR* circuit of Fig. 8.20 with $A = 1$, $B = 0$. For each gate, two lists *SF* and *L* are given below. *SF* is the super fault list of all faults, corresponding inputs and the output. *L* is the fault list i.e the list of faults which affect the output of the gate. *L* is a subset of *SF*. Referring to gate w, the possible faults are

Input A stuck at 0 (A_0)
Input A stuck at 1 (A_1)
Output w stuck at 0 (w_0)
Output w stuck at 1 (w_1)

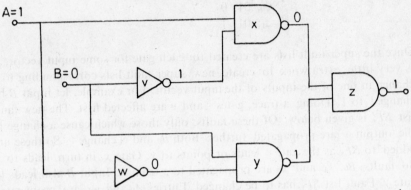

Fig. 8.20 Exclusive-*OR* circuit for which fault simulation has been demonstrated.

Of these faults, only A_0 and w_1 affect the output making it 1 instead of 0. So L_w contains only these two faults. *SF* and *L* for the other gates are similarly constructed. Note, for example, that though A is not a direct input to gate Y, A_0 is included in SF_y. This is because A_0 affects w and faults listed in L_w are included in SF_y.

Fault lists for $A = 1$, $B = 0$ in circuit of Fig. 8.20

Gate v

SF_v	L_v
B_1; 1; 0	B_1
v_0; 0; 0	v_0
B_0; 0; 1	
v_1; 0; 1	

Gate w

SF_w	L_w
A_0; 0; 1	A_0
w_1; 1; 1	w_1
A_1; 1; 0	
w_0; 1; 0	

Gate x

SF_x	L_x
A_0; 01; 1	A_0
v_0; 10; 1	v_0
B_1; 10; 1	B_1
x_1; 11; 1	x_1
A_1; 11; 0	
v_1; 11; 0	
x_0; 11; 0	

Gate y

SF_y	L_y
B_1; 10; 1	y_0
w_1; 01; 1	
A_0; 01; 1	
y_0; 00; 0	
B_0; 00; 1	
w_0; 00; 1	
y_1; 00; 1	

Gate z

SF_z	L_z
x_1; 11; 0	x_1
y_0; 00; 1	A_0
A_0; 11; 0	v_0
v_0; 11; 0	B_1
B_1; 11; 0	z_0
x_0; 01; 1	
y_1; 01; 1	
z_0; 01; 0	
z_1; 01; 1	

Once the super fault lists are created for each gate for some input vector, it is very little extra work to create new super fault lists corresponding to a change in one of the inputs of the input vector. For example, let input B be changed to 1. Doing a trace, gates v and y are affected first. The new fault list SF_v' is given below. Of these faults, only those which cause a change in the output v are propagated further. Both B_0 and v_1 change v. So these are added to SF_x' as the gate v leads or points to x. Gate x, in turn, leads to z. So faults B_0, v_1 and x_0 are propagated to z from x. Input B also leads to gate y. Fault list SF_y' has to be changed. Entries A_0 and w_1 are propagated to gate z. The new super fault lists and fault lists for $A = 1$, $B = 1$ are given below.

Fault lists for $A = 1$, $B = 1$ in circuit of Fig. 8.20

Gate v

SF_v'	L_v'
B_1; 1; 0	B_0
v_0; 0; 0	v_1
B_0; 0; 1	
v_1; 0; 1	

Gate w

SF_w'	L_w'
A_0; 0; 1	A_0
w_1; 1; 1	w_1
A_1; 1; 0	
w_0; 1; 0	

Gate x

SF_x'	L_x'
A_0; 00; 1	v_1
v_0; 10; 1	B_0
B_0; 11; 0	x_0
x_1; 10; 1	
A_1; 10; 1	
v_1; 11; 0	
x_0 10; 0	

Gate y

SF_y'	L_y'
B_1; 10; 1	y_0
w_1; 11; 0	A_0
A_0; 11; 0	w_1
y_0; 10; 0	
B_0; 00; 1	
w_0; 10; 1	
y_1; 10; 1	

Gate z

SF'_z	L'_z
x_0; 01; 1	x_0
y_0; 10; 1	A_0
A_0; 10; 1	v_1
v_1; 01; 1	B_0
B_0; 01; 1	w_1
y_1; 11; 0	z_1
z_0; 11; 0	y_0
z_1; 11; 1	
w_1; 10; 1	
x_1; 11; 0	

In a related technique for fault simulation, called deductive simulation, the fault lists L alone are constructed and stored. One has to do more work in finding the fault lists when any one input changes. However, the number of quantities to be stored is far less than in the case of concurrent simulation [5].

From the two input vectors ($A = 1$, $B = 0$) and ($A = 1$, $B = 1$), the faults observable at the output z are x_1, A_0, v_0, B_1, z_0, x_0, v_1, B_0, w_1, z_1 and y_0. The total number of faults possible for this circuit is 14. Of these, 11 have been detected by these two input vectors. We can say that the fault coverage for these two vectors together is 11/14 or 79%. The only faults not detected are A_1, w_0 and y_1.

Concurrent fault simulation requires a lot of memory because of all the super fault lists to be stored. Fault collapsing can be done to reduce the size of the lists. For example, any input of a NAND gate stuck at 0 is indistinguishable from the output of that gate stuck at 1. So these two faults can be collapsed into one as long as the stuck input is not an input to any other gate. In the example considered above x_0 and z_1 can be collapsed. Fault collapsing, in general, is far more difficult and the reader is directed to reference [5].

8.9 Compiler Driven Simulator for AHPL

A different kind of simulator is discussed in this section. A language called AHPL (A Hardware Programming Language) developed and described by Hill and Peterson [2] is used to describe the digital system. The language also describes the control sequences, and asynchronous and conditional transfers. AHPL is a register transfer language (RTL) and especially useful for microprogram based architectures. AHPL uses notational conventions of APL (A Programming Language) and only those APL operates have been employed in it that can be readily interpreted as hardware primitives. The language itself is not described here and the reader is directed to Ref. [2].

At IIT Kanpur, a compiler has been developed for AHPL [9]. This involves taking an AHPL program as a source code and producing a target code to simulate the running and behaviour of the digital system described by AHPL. Hence, the target code when executed will present to the user a view of the system running—executing the control sequence in successive clock cycles.

The actions in a clock cycle would include transfers to and from registers, loading from and onto lines, and branching (conditionally or otherwise) to some other cycle sequence. Unlike a gate level simulator the description is at a functional level. No delays can be simulated. The simulation is not event directed. Fault simulation is not possible. However, the simulator allows the user to monitor values of lines, registers, etc. interactively.

The compiler functions in two parts. First the AHPL code is converted to PASCAL. Then the standard PASCAL compiler is used to compile the PASCAL program generated. Not only is the need to compile AHPL code avoided, but parts of the digital system can be described in PASCAL. Later, these can be merged with the converted AHPL code. It is found that the input/output routines are described more conveniently in PASCAL.

8.10 Summary

Simulation at the logic level has been described in this chapter. The emphasis has been on a gate level simulation. Execution speeds can be considerably improved by selectively analysing only those gates which lie on the signal flow path. Timing information can be obtained by assigning different delays to different gates. Logic simulation is related to fault simulation and fault simulators usually have routines for logic simulation. Logic simulators must have special models for unidirectional and bidirectional pass transistors to handle MOS logic circuits. A logic simulator developed at IIT Kanpur was described.

References

1. Y. Chu, Computer Organisation and Microprogramming, Chap. 1 Prentice-Hall Inc, 1972.
2. F.J. Hill and G.R. Peterson, Digital Systems: Hardware Organisation and Design, Chap. 5, John Wiley and Sons, 1973.
3. 'LAMP', Bell System Technical Journal, pp. 1442-1475, Oct. 1974.
4. A. Miczo, Digital Logic Testing and Simulation, Harper and Row, Publishers, 1986.
5. M.A. Breuer and A.D. Friedman, Diagnosis and Reliable Design of Digital Systems, Chap. 4, Computer Science Press, 1976.
6. E. Horowitz and S. Sahni, Fundamentals of Data Structures, Chap. 4, CBS Publishers and Distributors, 1983.
7. R.S. Raghunandan, SIMLOG: A Logic Simulator, M. Tech. Thesis, Dept. of Electrical Engg., IIT Kanpur, 1987.
8. E.B. Eichelberger, Hazard Detection in Combinational and Sequential Switching Circuits, IBM J. Res. Development, vol. 9, pp. 90-99, March, 1965.
9. R. Chandra, S. Khuller and P.S. Sahai, A Compiler for AHPL, B. Tech. Project Report, Dept. of Computer Science and Engineering, IIT Kanpur, 1986.
10. R.M. McDermott, Transmission Gate Modelling in an Existing Three-value Simulator, 19th Design Automation Conference, pp. 678-681, 1982.

11. Bipolar Microprocessor Logic and Interface, AM 2900 Family, 1985, (Data Book).
12. C. Mead and L. Conway, Introduction to VLSI Systems, Chapter 6, Addison-Wesley Publishing Co., 1980.

Problems

1. Consider a 2-bit ring counter with two master-slave *JK* flipflops used as *D* flipflops as shown in Fig. P 8.1. If the initial state is 1-0 analyse the circuit for one complete clock cycle using selective trace. Assume all gates have the same delay.

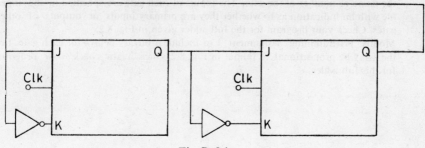

Fig. P. 8.1

2. (a) For the parity check logic circuit shown in Fig. P 8.2 assume that all gate outputs are 'don't know' or state 2 initially. If an input $A = 1$, $B = 1$, $C = 0$, $D = 1$ is given, do the initial state analysis of the circuit. The three exclusive-*OR* blocks should first be replaced by the circuit of Fig. 8.4.
   (b) The input C is now changed from 0 to 1. Analyse the 'transient response'-assuming all gates to have the same delay. Do the analysis with and without the selective trace feature.

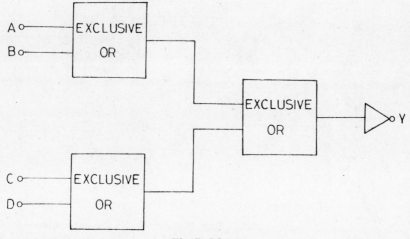

Fig. P. 8.2

3. For the circuit of Fig. P 8.2, construct the tables showing the interconnections. Do this for each of the schemes given in Section 8.5.
4. Repeat problem 1 assuming the delay of each gate to be equal to the fanout.
5. Assume that the fault lists for the circuit in Fig. 8.20 are available for $A = 1$, $B = 0$ as given in the text. Let input A be changed to 0. Construct the new fault

lists from the old ones tracing the signal flow path. Check your result with that computed from scratch.
6. Find the conditions under which a NOR latch may lead to a race.

Programming Assignments

1. Write a program to analyse combinatorial circuits only containing 2 input NAND gates. Assume all gates to have the same propagation delay. Each gate is specified by an index number and name. The output of each gate is referred to by the name or number of the gate. The inputs of each gate are specified in the input file with an indication as to whether they are primary inputs or outputs of other gates. Check your program for the full adder given in Fig. 8.2.
2. Modify programming assignment 1 to include arbitrary delays for each gate. Let the delay be proportional to fanout in Fig. 8.2. Once again check your program for this full adder.

9

Relaxation Based Methods for Transient Analysis

Full fledged circuit simulation takes up considerable CPU time, especially for nonlinear transient analysis. For example, a 700 MOSFET circuit took over 4 hours of CPU time on a VAX 11/780 for 200 time steps [1]. Various techniques have been tried to improve the speed of SPICE-like circuit simulators. Among these have been

(a) Use of specialised hardware called engines. SPICE engines which run upto 100 times faster have been reported [2].

(b) Use better data structures to improve the execution speed of existing programs.

(c) Vector processing with supercomputers like CRAY.

(d) Parallel processing (at subroutine level) on multiprocessor systems [13].

A second approach, which is the subject of discussion here, is to use special purpose simulators. Of special interest are simulators for MOS digital circuits. A very large number of ICs fall in this category and a variety of simulators have been built for analysing them. Special properties of MOS digital circuits are exploited. The algorithms may or may not be suited to general purpose simulators like SPICE. Most of them are based on some relaxation kind of numerical technique [1]. Some of them go by the name of timing simulators. This is because they are used mainly for transient analysis i.e. to provide timing information ([3], [4]).

MOS digital circuits have several features which can be taken advantage of in designing a special purpose simulator for non-linear transient analysis. First, except where pass transistors are used, no DC coupling exists between gates. The loading on each gate is capacitive. Many features of logic simulators can be used. Simulators can be made event driven and a data structure like that used in Section 8.5 can be used to trace the signal flow path. Inactive gates need not be analysed. MOS transistors occur in limited configurations. For NMOS (or PMOS) circuits, these are the load transistor, driver transistor and pass transistor. As in the first timing simulator MOTIS [3], look up tables can be used to store the currents through the transistors occurring in these three configurations. This considerably speeds up execution at the expense of some memory. Finally, as feedback paths are limited, iterative relaxation based techniques converge fast.

9.1 Basic Theory Behind Relaxation Based Methods

Let it be necessary to solve m coupled equations in m unknowns. These equations may be differential or algebraic, linear or non-linear. In general, we have two possible methods of solution—direct and iterative. An example of a direct method is Gaussian elimination. Among the iterative techniques, we have relaxation methods and non-relaxation methods. In a relaxation method, the unknowns are decoupled. We solve for one unknown at a time leaving the others fixed. Assume, for example, that we are solving a system of m simultaneous linear equations. We make an initial guess for all n variables. We then refine this guess by solving the ith equation for variable i holding the other variables fixed at the initial guess. Once we have done one round for all n unknowns, we iterate a second time. This is continued till convergence.

The decoupling of the unknowns can be done at any level. Figure 9.1 shows various alternatives for solving non-linear differential equations. We can apply relaxation methods at the first level itself (right-most branch). We solve the ith differential equation for the ith unknown (assuming the differential equations are suitably ordered) holding the other unknowns fixed. This, of course, must be done iteratively till convergence. Another possibility is to first reduce the non-linear differential equations to non-linear algebraic or difference equations. At the non-linear equation level, we decouple the unknowns. We have, then, m non-linear equations, each in one unknown. These can be solved iteratively till convergence. This corresponds to the middle branch in Fig. 9.1. A circuit simulator like SPICE corresponds to the branch shown with double lines.

Circuit simulation programs have been written following the various strategies outlined above. Often some combination is used. If relaxation methods are used, then decoupling (i.e. applying relaxation methods) is done at either the differential equation level, non-linear equation level or the linear equation level. The term relaxation comes from the fact that we are said to 'relax' one variable at a time, holding the other constant. Whatever be the level at which relaxation is applied, two common relaxation techniques used are the Gauss-Seidel and Gauss-Jacobi (G-S and G-J respectively) techniques. Gauss-Seidel is a special case of the more general successive over-relaxation (SOR) technique. These are described in the next section. Relaxation methods are particularly attractive for MOS digital circuits because individual elements are to a large extent decoupled. At one extreme we have logic simulation where all elements are totally decoupled and at the other we have bipolar analog circuits where all elements are very tightly coupled. MOS digital circuits fall somewhere in between. We can therefore hope to have some measure of success in applying relaxation methods to such circuits.

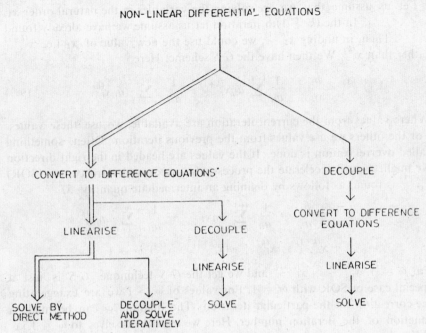

Fig. 9.1   Various possible approaches to solve non-linear differential equations

9.2   Gauss-Seidel and Gauss-Jacobi as Applied to Linear Equations

The G-S and G-J techniques are discussed in many standard numerical analysis text books ([6], [7], [8]). Often they are applied to solve linear equations and we will discuss their application to linear equations first. Let the system of equations be of the form $[A] [x] = [B]$. Rather than use a direct method like Gaussian elimination, we will see the pros and cons of using an iterative scheme like G-J or G-S. Both schemes require an initial guess $[x^{(0)}] = [x_1^{(0)}, x_2^{(0)}, \ldots x_m^{(0)}]$. If we assume that the diagonal is strong, then the value of the ith variable is mainly governed by the ith equation.

In the G-J technique we use the ith equation to find $x_i^{(1)}$ from $x_i^{(0)}$ holding the other variables at $x^{(0)}$.

$$x_i^{(1)} = \frac{b_i}{a_{ii}} - \frac{1}{a_{ii}} \sum_{\substack{j=1 \\ j \neq i}}^{m} a_{ij} x_j^{(0)} \tag{9.1}$$

After we have found all $x_i^{(1)}$ we find $x_i^{(2)}$ and so on till convergence. In general, the $(k + 1)$th iterate is found from the kth iterate as

$$x_i^{(k+1)} = \frac{b_i}{a_{ii}} - \frac{1}{a_{ii}} \sum_{\substack{j=1 \\ j \neq i}}^{m} a_{ij} x_j^{(k)} \tag{9.2}$$

An interesting feature of the G-J technique is that the order in which the variables are solved is immaterial. This is not true of the G-S technique given below.

Let us assume that we are solving the variables in the natural order x_1, $x_2, \ldots, x_m$. In the $(k + 1)$th iteration let us assume we have already found $x_1^{(k+1)}$. Then, in finding $x_2^{(k+1)}$, we could use the new value of x_1 i.e. $x_1^{(k+1)}$ rather than $x_1^{(k)}$. We then have the G-S scheme. Here,

$$x_i^{(k+1)} = \frac{b_i}{a_{ii}} - \frac{1}{a_{ii}} \sum_{j=1}^{i-1} a_{ij}x_j^{(k+1)} - \frac{1}{a_{ii}} \sum_{j=i+1}^{m} a_{ij}x_j^{(k)} \qquad (9.3)$$

Where values from the current iteration are available, we use these values. For the others we use values from the previous iteration. Often, something called overrelaxation is done. If the values are headed in the right direction we might as well accelerate the process. In successive overrelaxation (SOR) $x_i^{(k+1)}$ is found as follows by defining an intermediate quantity $\bar{x}_i^{(k+1)}$.

$$\bar{x}_i^{(k+1)} = \frac{b_i}{a_{ii}} - \frac{1}{a_{ii}} \sum_{j=1}^{i-1} a_{ij}x_j^{(k+1)} - \frac{1}{a_{ii}} \sum_{j=i+1}^{m} a_{ij}x_j^{(k)}$$

$$x_i^{(k+1)} = x_i^{(k)} + \omega(\bar{x}_i^{(k+1)} - x_i^{(k)}), \omega \geqslant 1$$

For $\omega = 1$, $\bar{x}_i^{(k+1)} = x_i^{(k+1)}$ and we get the G-S technique. G-S is just a special case of SOR with $\omega = 1$. For values of $\omega > 1$ we are exaggerating the correction for the particular iteration. The quantity ω can be made a function of the iteration number. Here we restrict ourselves to $\omega = 1$ i.e. G-S technique.

In general, G-S techniques converge faster than G-J techniques. In the case of a lower triangular matrix for $[A]$, G-S converges in one iteration while G-J takes n iterations. The order of computation in G-S is important and could affect the time for solution drastically. As an extreme case, if $[A]$ is lower triangular and one solves in the order $x_m, x_{m-1} \ldots x_1$, the solution takes m iterations instead of one. Often a knowledge of the problem may help in the ordering. If we are studying the response to an input change, the order would naturally be the one in which various node voltages occur as one traces the signal flow path.

In short, the basic difference between G-S and G-J techniques is that in G-S we use values from the current iteration wherever possible. We do not in G-J. Naturally, the order of computation is important in G-S but not in G-J.

The question arises as to where one should use an iterative technique as opposed to a direct technique like Gaussian elimination. For G-J or G-S each x_i in an iteration requires m operations. The term b_i/a_{ii} is common to all iterations. Each iteration for m variables takes m^2 operations. If we need k iterations we have km^2 operations in all. For Gaussian elimination we need $m^3/3$ operations (ignoring m^2, m terms). This simple analysis tells us that as long as k is less than $m/3$, G-J/G-S techniques are faster. This does not take sparsity into account. A comparison is difficult for sparse matrices because the number of operations cannot be easily calculated in the case of direct techniques. Assume that on the average, every row contains p nonzero elements. Then G-S/G-J require knp operations. For LU decomposition, assume that no fills occur. One would expect pm^2/a operations for LU decomposition where a would depend on the matrix.

In the various relaxation based simulators decoupling is not done at the linear equation level. It is done at the non-linear equation level or at the differential equation level. These are discussed in the following sections.

9.3 Gauss-Seidel and Gauss-Jacobi Applied to Non-Linear Equations

Consider a set of non-linear equations of the form

$$f_1(x_1, x_2, \ldots x_m) = 0$$
$$f_2(x_1, x_2, \ldots x_m) = 0$$
$$\vdots$$
$$f_m(x_1, x_2, \ldots x_m) = 0$$

These can be decoupled at the non-linear equation level using G-S or G-J techniques. Let the initial guess be $[x^{(0)}] = (x_1^{(0)}, x_2^{(0)}, \ldots x_m^{(0)})$. To get $x_i^{(1)}$ using G-S we solve

$$f_i(x_1^{(1)}, x_2^{(1)}, \ldots x_{i-1}^{1}, x_i^{(1)}, x_{i+1}^{(0)}, \ldots x^{(0)}) = 0 \quad \text{for} \quad x_i^{(1)}$$

Using G-J we solve

$$f(x_1^{(0)}, x_2^{(0)}, \ldots x_i^{(1)}, \ldots x_m^{(0)}) = 0 \quad \text{for} \quad x_i^{(1)}$$

In either case we are solving non-linear equations in just one unknown. Other techniques like the secant method are available to solve such equations. Of course, many iterations have to be done before the solution converges.

Let us assume, for the moment, that we decouple the unknowns using Gauss-Seidel and then solve the non-linear equations in one unknown using Newton-Raphson. We therefore have two levels of iteration—an outer Seidel iteration and an inner Newton iteration. We will refer to this method of solving non-linear equations as the Seidel-Newton method. In every Seidel iteration we solve m non-linear equations which are decoupled. Each of these equations is solved in many Newton-Raphson iterations. Both inner and outer loop iterations must be carried to convergence, in principle. However, it is possible to trade off some Newton iterations for Seidel iterations. One could, for example, do just one Newton iteration but carry the outer Seidel iterations to convergence.

Another possibility is to first apply Newton-Raphson to the system of non-linear equations. These are then linearised. These linear equations can be solved using, say, Gauss-Seidel. We will refer to this method as Newton-Seidel. Again one can do more Newton iterations and reduce the number of Seidel iterations. Most relaxation based programs use Seidel-Newton (or

Jacobi-Newton) as opposed to Newton Seidel. Often, the inner loop has just one iteration. Seidel-Newton is preferred because

(a) If k_s Seidel iterations and k_n Newton iterations are needed, then the total solution using Seidel-Newton requires mk_sk_n derivatives. Newton-Seidel requires m^2k_n derivatives. As each f_i is a function of just 2 or 3 voltages, the number is more like amk_n where 'a' is less than 5. As mentioned earlier, the outer loop is often carried to convergence with just one inner loop iteration. In such a situation, Newton-Seidel requires about 'a' times more derivatives. In many timing simulators both k_n and k_s are 1. Then again, Newton-Seidel requires more derivatives.

(b) Convergence of Newton-Raphson is quadratic, but that of G-J or G-S is linear. In other words, Newton-Raphson converges much faster in the region of the solution. One could therefore get away with fewer Newton iterations than G-S or G-J iterations. As we would like to do fewer (may be just one) iterations in the inner loop, Newton-Raphson is a better choice for the inner loop.

(c) Once the non-linear equations are decoupled, they may be simple enough that they can be solved directly, rather than by using Newton-Raphson.

A large number of programs are based on decoupling of equations at the non-linear level using Gauss-Seidel or Gauss-Jacobi techniques. These go by the general name of 'timing simulators' because they give timing information. In other words they do transient analysis. The non-linear differential equations are reduced to difference equations by using a technique like Backward Euler. The associated network models could be used to replace the capacitors to get a nonlinear DC network at each time step. This is solved using relaxation techniques. Consider a typical node like A shown in Fig. 9.2. The node equation at A gives

$$C_{AA}\frac{dv_A}{dt} + C_{A1}\frac{d}{dt}(v_A - v_1) + C_{A2}\frac{d}{dt}(v_A - v_2)$$

$$-i_1(v_A) + i_2(v_A, v_1) + i_3(v_A, v_2, v_3) = 0$$

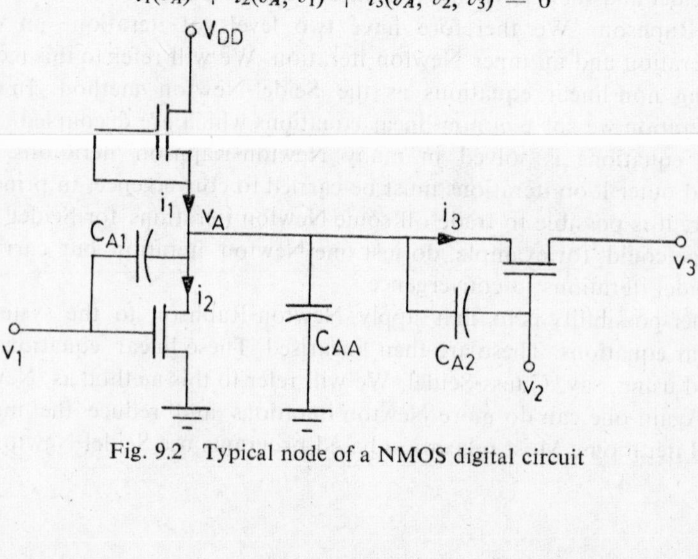

Fig. 9.2   Typical node of a NMOS digital circuit

Rewriting, we get

$$\frac{dv_A}{dt}(C_{AA} + C_{A1} + C_{A2}) - C_{A1}\frac{dv_1}{dt} - C_{A2}\frac{dv_2}{dt}$$

$$= g_A(v_A, v_1, v_2, v_3) \tag{9.4}$$

where $\qquad q_A = i_1 - i_2 - i_3$

For a MOS digital circuit the set of node equations can be written in the following general form

$$[C][\dot{v}] = [g([u], [v])] \tag{9.5}$$

where $[C]$ is the capacitance matrix, $[\dot{v}]$ the vector of node voltage time derivatives and $[g]$ the vector of functions which depend on the various node voltages. The diagonal elements C_{ii} in $[C]$ represent the total capacitance connected to node i. The non-diagonal elements are all negative and $|C_{ij}|$ represents the capacitance between node i and node j. The function g_i at node i represents the net current flowing into node i apart from the capacitor currents. If Eq. (9.4) were represented in the form of Eq. (9.5) then C_A, the diagonal term at node A, would be equal to $(C_{AA} + C_{A1} + C_{A2})$. $[u]$ is the vector of known input voltages.

Let us apply the Backward-Euler technique to Eq. (9.4). We get at time t_{n+1},

$$C_A\frac{(v_{An+1} - v_{An})}{\Delta t} - \frac{C_{A1}}{\Delta t}(v_{1n+1} - v_{1n}) - \frac{C_{A2}}{\Delta t}(v_{2n+1} - v_{2n})$$

$$= g_A(v_{1n+1}, v_{An+1}, v_{2n+1}, v_{3n+1}) \tag{9.6}$$

There will be similar equations at the various other nodes, all of them now reduced to non-linear algebraic equations. The unknowns can be decoupled using G-S. Let the order of computation be . . . 1, A, 2 and 3, ... We will assume initial guesses $v_{1n+1}^{(0)}$, $v_{An+1}^{(0)}$, $v_{2n+1}^{(0)}$ and $v_{3n+1}^{(0)}$. Then the first iteration value $v_{An+1}^{(1)}$ is given by

$$C_{AA}\frac{(v_{An+1}^{(1)} - v_{An})}{\Delta t} - \frac{C_{A1}}{\Delta t}(v_{1n+1}^{(1)} - v_{1n}) - \frac{C_{A2}}{\Delta t}(v_{2n+1}^{(0)} - v_{2n})$$

$$= g_A(v_{1n+1}^{(1)}, v_{An+1}^{(1)}, v_{2n+1}^{(0)}, v_{3n+1}^{(0)}) \tag{9.7}$$

From the G-S order of computation, it is clear that $v_{1n+1}^{(1)}$ would be computed before $v_{An+1}^{(1)}$. The above non-linear equation in one unknown, namely $v_{An+1}^{(1)}$, may be solved directly or by any iterative method like Newton-Raphson.

In a class of simulators called Iterated Timing Analysis simulators (ITA simulators) several G-S iterations are done till convergence is reached [1]. This approach is quite rigorous but time consuming. Table 9.1 compares execution times and memory for ITA with SPICE. While ITA is clearly superior for the digital circuit, its performance is worse for the analog circuit. In an effort to save computer time, just one G-S iteration is done in many simulators, [4]. These simulators, called timing simulators, definitely

Table 9.1   Comparison of SPICE with timing and ITA simulators [1].

Circuit :		ENCODER/DECODER		OPAMP	
MOSFETS :		1326		15	
NODES :		553		14	
	TIME (sec)	MEMORY (K bytes)	TIME (sec)	MEMORY (K bytes)	
SPICE 2G	115,840	2,420	59.8	29.0	
SPLICE 1.6(ITA)	1,740	68.9	114.3	9.3	
SPLICE 1.3 (TIMING)	789	64.4	—	—	

cannot be applied to analog circuits. Under certain conditions, they give good results for MOS digital circuits. Table 9.1 shows that the timing simulator SPLICE 1.3 is faster than both SPICE and the ITA simulator SPLICE 1.6[1]. In most timing simulators, the decoupled non-linear equations like Eq. (9.7) are solved doing just one Newton-Raphson iteration. This can also lead to problems.

A major problem with timing simulators is that they cannot handle floating capacitors. In Eq. (9.7), the natural choice for the initial guess $v_{1n+1}^{(0)}$, $v_{An+1}^{(0)}$, etc., are the values v_{1n}, v_{An}, etc. from the previous time step. Making these substitutions, Eq. (9.7) reduces to

$$\frac{C_A}{\Delta t}(v_{An+1}^{(1)} - v_{An}) - \frac{C_{A1}}{\Delta t}(v_{1n+1}^{(1)} - v_{1n}) - \frac{C_{A2}}{\Delta t}(v_{2n} - v_{2n})$$

$$= g_A(v_{1n+1}^{(1)}, v_{An+1}^{(1)}, v_{2n}, v_{3n}) \qquad (9.8)$$

The third term in the above equation goes to zero. In other words, floating capacitors connected to nodes not yet computed are treated as grounded capacitors. If G-J had been used instead of G-S, the second term would also have gone to zero. In that case, all floating capacitors connected to a node would have been treated as grounded capacitors. Timing simulators also have problems with feedback paths. In Fig. 9.3, the gates are evaluated in the sequence 1, 2, 3. When evaluating gate 1, the output of gate 3 is not yet available. So the value from the previous time step is used. There is an error equal to one time step in such situation. If the time step is small, the error may not be serious. Similarly, in Fig. 9.2, v_2 and v_3 affect v_A. As v_2 and v_3 are computed after v_A, an error of one time step will occur.

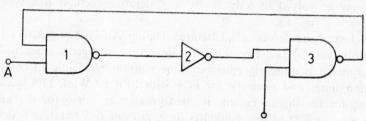

Fig. 9.3   Digital circuit with feedback

The next question that arises relates to how one solves the decoupled non-linear equations like Eq. (9.8). Newton-Raphson is the natural choice. Timing simulators usually do just one Newton-Raphson iteration. On the whole one combined Seidel-Newton or Jacobi-Newton iteration is done. The initial Newton guess is also the value from the previous time step. It was pointed out in the last paragraph that floating capacitors are not properly handled by single iteration Seidel or Jacobi techniques. For this reason, floating capacitors are converted to equivalent grounded capacitors. If this is done, the general form of the decoupled non-linear equation at the ith node is

$$\frac{C_{ii}}{\Delta t}(v_{in+1}^{(1)} - v_{in}) = g_i([u_{n+1}], v_{1n+1}^{(1)}, v_{2n+1}^{(1)} \ldots v_{in+1}^{(1)}, v_{1n} \ldots v_{mn})$$

where $[u]$ = vector of input voltages, m = number of unknown node voltages and n, $n + 1$ refer to time t_n and t_{n+1} respectively. Applying $N-R$ to the above equation we get (for an initial guess $v_{in+1}^{(10)} = v_{in}$)

$$-g_i([u_{n+1}], v_{1n+1}^{(1)}, v_{2n+1}^{(1)}, \ldots v_{in}, v_{i+1n} \ldots v_{mn}) +$$

$$(v_{in+1}^{(11)} - v_{in})\frac{C_{ii}}{\Delta t} - g_i'([u_{n+1}], v_{1n+1}^{(1)}, v_{2n+1}^{(1)} \ldots v_{in}, v_{i+1n}, v_{mn}) = 0$$

The term v_{in+1} has two superscripts. The first refers to the Seidel iteration number and the second to the Newton iteration number. g_i' is the derivative of g_i with respect to $v_{in+1}^{(1)}$. If we do just one Seidel and one Newton iteration $v_{in+1}^{(1)} = v_{in+1}^{(11)} = v_{in+1}$.

$$v_{in+1} - v_{in} = \Delta v_i = \frac{g_i([u_{n+1}], v_{1n+1}, v_{2+1} \ldots v_{in}, v_{i+1n} \ldots v_{mn})}{\dfrac{C_{ii}}{\Delta t} - g_i'([u_{n+1}], v_{in+1}, v_{2n+1} \ldots v_{in}, v_{i+1n} \ldots v_{mn})}$$

$$(9.8)$$

This is the basic equation used in timing simulators based on a one sweep Seidel-Newton iteration. For Jacobi-Newton, the values of all node voltages would be those from the previous time step t_n. The function g_i represents the net current flowing into the grounded capacitance C_{ii} and g_i' represents its derivative with respect to v_i. The derivative g_i' also represents the sum of the dynamic conductances at node i. Simulators differ in the way g_i and g_i' are calculated. The first timing simulator, MOTIS [3], used the Jacobi-Newton technique. Functions g_i were computed from look-up tables for the driver, load and pass transistor currents. Conductances g_i' were evaluated approximately from asymptotic values. In MOSIMR [9] (see next section), both g and g' are evaluated directly from the expressions. But the simple Schichman-Hodges model [10] for the MOSFET is used.

One may wonder how useful timing simulators can be with all their inherent limitations and the approximations used. One reason they work is that many of these limitations disappear with a small time step. If Δt is small, the voltages cannot change significantly in a time step. Therefore, one can arrive at a good initial guess just by using values from the previous time step. It is for this reason that we are able to get away with one combined Seidel-Newton iteration. Time steps used in timing simulators may be one-tenth of the value used in a proper circuit simulator like SPICE. ITA simulators do not suffer from the limitations of timing simulators. For some applications, they may be the best compromise between timing and circuit simulators.

Relaxation based simulators for digital circuits often borrow concepts used in connection with logic simulators. Dormant parts of the circuit are not analysed. Usually they are event driven. In response to an input change, the signal flow path is traced. Only those elements lying in the path are analysed. A linked list structure must be used to do the trace. A scheduler is also necessary. However, we monitor the gates/elements continuously. In a logic simulator the output can change only 'd' time steps after the input changes, where d is the delay. In a timing simulator the output begins to change immediately. When any input changes, all elements in the signal flow path must be analysed for all subsequent time steps till all outputs stabilise. The simulator MOSIMR, described in the next section, improves on this by considering only those input changes as significant which are greater than a user specified value. Seidel methods have an important advantage over Jacobi methods in the way an input change propagates. In Fig. 9.3 let input A change. Then, if the gates are analysed in the order (1, 2, 3) gate 3 shows a change in the same time step. If a Jacobi scheme is used, then gate 3 will have an output change three time steps later. In a chain of gates, a change propagates gate by gate for Jacobi methods, but immediately for Seidel methods. Of course, this is true only if gates are analysed in the order they occur along the signal flow path. As pointed out earlier, the order of computation is important for Seidel methods.

Floating capacitors can be taken into account even in the one sweep timing simulators by proper partitioning of the circuit. A block like that in Fig. 9.2 can be considered as one super element. At each time step this big decoupled element is solved rigorously using several N-R iterations. The bigger partitioned elements are solved in sequence according to either a Seidel or a Jacobi technique.

9.4 Example of an Event Driven Timing Simulator

Timing and power dissipation are two important performance indices of a MOS digital circuit. With this in view a program to simulate NMOS circuits called MOSIMR has been developed at IIT Kanpur. It provides the user with initial logic states as well as timing waveforms of node voltages and element power over the simulation period.

The focus is on minimising CPU time by exploiting latency wherever possible. Any input change is traced through the circuit and only those elements lying on the signal flow path are analysed. A linked data structure of the circuit topology is constructed to order the processing of nodes in accordance with the signal flow. Partitioning of the circuit into independent subsystems is implemented by constructing fan-in/fan-out tables for the load nodes. A pass transistor is both a fan-in and fan-out element to its source and drain terminals. The circuit trace or node computation sequence is determined by a circular queue with two pointers which acts as a scheduler. If the rear pointer overtakes the front pointer, the queue is recognised as full and simulation is terminated. When the front overtakes the rear the trace comes to an end.

A change in any input takes a finite time to propagate through various devices on the signal flow path. Even if a node lies on the path, it may remain latent at a particular time step. This may happen either if the change has not propagated to the node or if the change has taken place fully. MOSIMR exploits this latency. Thus, within a selected path, only the active nodes are analysed. A flag associated with each node indicates whether it is active or not. A user specified Δ is compared with the change ϵ in node voltage. A change is recognised only if Δ is less than ϵ. Of course, the accuracy of computation depends on the choice of Δ. Where an approximate initial analysis is desired Δ can be large. Exploitation of signal latency is naturally more effective when the chain in the signal flow path is longer.

The actual computation is done as in most timing simulators. Nodal equations of the form given in Eq. (9.5) are to be solved with the initial condition $[v(0)] = [v_0]$. The Backward-Euler integration scheme is used to discretise the differential equations, the Gauss-Seidel technique to decouple the unknowns and the Newton-Raphson method to solve the resulting non-linear equations. Actually, for simplicity, just one combined Seidel Gauss-Newton iteration is done. For this reason MOSIMR reduces all floating capacitances to equivalent grounded capacitances. At each node, path-length stray capacitances are computed from the width, length and information as to whether the path is metal as poly. The Schichman-Hodges model [10] is used to describe the NMOS transistors.

MOSIMR is a mixed-mode simulator in the way initialisation is done. First a logic simulation is done. Then actual voltages and currents are computed from the parameters of the transistors used for the various gates. Floating nodes associated with outputs of 'OFF' pass transistors are assumed to be logical at '0'.

MOSIMR has been written in FORTRAN-10 and implemented on the DEC 10/90 system. The program is over 3000 lines. The user describes the circuit in terms of logic gates, latches and pass transistors. Functional blocks can be defined as sub-circuits. Node and element numberings within the sub-circuit description are local. The circuit description module offsets local node and element numbers replaces dummy external nodes with actual node numbers and sets up links with the main circuit. MOSIMR is divided into six functional modules as follows:

1. Circuit description
2. Initialisation
3. Input

4. Analysis
5. Output
6. Control

MOSIMR was used to analyse several circuits and the results were compared with those from SPICE 2G.6. In each case there was no noticeable loss of accuracy, but execution times were considerably faster. In addition MOSIMR was able to provide power dissipation values. Table 9.2 compares execution times for six sample circuits.

Table 9.2   Comparison of MOSIMR and SPICE 2G.6

Circuit No.	1	2	3	4	5	6
No. of nodes	4	7	4	13	20	50
No. of MOS devices	6	6	7	21	52	82*
CPU-SPICE(s)	6.45	5.18	6.06	18.5	49.66	55.24
CPU-MOSIMR(s)	1.58	1.08	0.78	1.91	3.65	2.29
CPU SPICE/ MOSIMR	4.08	4.8	7.77	9.69	13.61	24.12

*SPICE input had only 58 transistors, as inverters were used instead of inverting buffers.

Circuit 1: NAND latch
Circuit 2: Two phase dynamic shift register cell
Circuit 3: MOS Ring Oscillator
Circuit 4: One-bit full adder
Circuit 5: Two-bit Johnson Ring Counter
Circuit 6: Circulating refresh register with read/write buffers.

In general, larger circuits give more favourable results for MOSIMR as latency is better exploited. For circuit 6, the latency in signal propagation was better exploited as there was a relatively long chain of 15 gates and pass transistors through 3 shift register cells and the R/W logic block.

Figure 9.4 shows the 2-bit Johnson ring counter analysed. Figure 9.5 gives the MOSIMR output and Fig. 9.6 the SPICE output. Analysis starts from the reset state with $R = 1$. Thereafter the reset is removed by making $R = 0$ and one cycle of the 2-phase clock is applied at nodes 1 and 2 to effect transition to the next state. MOSIMR results agree very well with SPICE. Figure 9.7 shows the plot of total power dissipation against time.

The special purpose timing simulator has been found to be comparable in performance to general purpose simulators like SPICE, while being considerably faster. The decoupling of the nodes by diagonalizing [C] makes MOSIMR O(n) compared to SPICE which is approximately O($n^{1.3}$ to $n^{1.5}$) for n nodes [1]. Additional enhancement in speed is due to selective trace and signal wave profile detection to weed out latent elements. Several features of logic simulators have been used and the initialisation routine does only logic simulation in its first sweep. MOSIMR is more accurate than

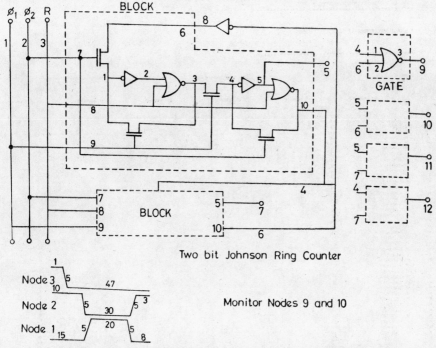

Fig. 9.4   Two bit Johnson ring counter analysed by MOSIMR

most timing simulators as the actual MOS transistor equations are used to calculate currents and dynamic conductances. The accuracy is, however, dependent on the proper choice of time step.

9.5 Waveform Relaxation 11

Referring to Fig. 9.1, we have so far not discussed solution techniques corresponding to the rightmost branch. This involves decoupling the unknowns at the differential equation level. That is, we solve for $v_1(t)$ holding other waveforms (as a function of time) fixed. We relax $v_1(t)$ first, then $v_2(t)$ and so on. Rather than the value of a variable at a particular time, the whole waveform $v(t)$ is relaxed. Hence, the name waveform relaxation. The matrix of values given below shows another way of looking at waveform relaxation.

$$u_1 u_2 \ldots u_r : v_1 v_2 \ldots v_m$$

t_1		
t_2		
t_3	KNOWN	UNKNOWN
.		
.		
.		
t_n		

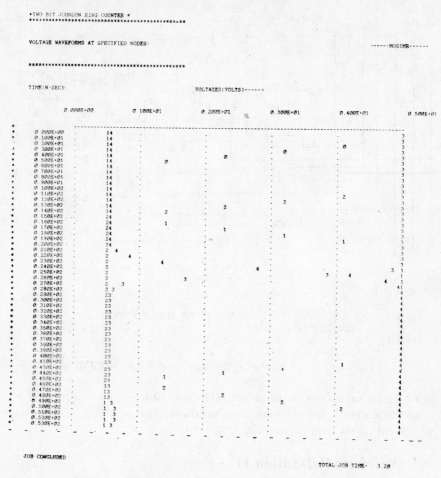

Fig. 9.5 MOSIMR output for Fig. 9.4. Legend is as follows
0: V(3); 1: V(1); 2: V(2); 3: V(9); 4: V(10)

What one has to finally solve for are the values of m node voltages at the n time intervals. The values of the r inputs, u_1 and u_r, are known at all times. In timing simulators, ITA simulators and circuit simulators we calculate row wise. First all voltages are calculated at t_1, then at t_2 and so on till t_n. In waveform relaxation, we proceed columnwise. First $v_1(t)$ is calculated at all times. Then we calculate waveforms $v_2(t)$, and so on till $v_m(t)$. The process is iterated till convergence is reached.

Here again one can talk of Seidel and Jacobi techniques. In both, we have to make an initial guess for all voltages at all times i.e. for the entire $(m \times n)$ matrix of values. The initial voltages at all m nodes at $t=0$ would be given. In the absence of any other information, the values at $t=0$ can be assumed at all times as an initial guess. Let the initial guess for the waveforms be $v_1^{(0)}(t)$, $v_2^{(0)}(t)$, ... $v_m^{(0)}(t)$. When using the G-S technique to decouple the waveforms, waveforms from the current iteration are used if already calcu-

```
°1W2 BIT JOHNSON COUNTER*
^1*1*            TRANSIENT ANALYSIS                    TEMPERATURE =  27.000 DEG C

***********************************************************************************************************************

^LEGEND:

1: V(3)
+: V(1)
*: V(2)
$: V(9)
0: V(10)


I
   TIME      V(3)

 (*1=$0)------------  0.000D+00           2.000D+00           4.000D+00           6.000D+00           8.000D+00
                     .-------------------------------------------------------------------------------------------
 0.000D+00  5.000D+00 .       +0        .                   .                I  .
 1.000D-09  5.000D+00 .       +0        .                   .                I  .
 2.000D-09  4.100D+00 .       +0        .                 .I                 I  .
 3.000D-09  3.200D+00 .       +0        .            I     .                 I  .
 4.000D-09  2.300D+00 .       +0        .       I         .                  I  .
 5.000D-09  1.400D+00 .       +0      I .                 .                  I  .
 6.000D-09  5.000D-01 .       I0       .                  .                  I  .
 7.000D-09  5.000D-01 .       I0       .                  .                  I  .
 8.000D-09  5.000D-01 .       I0       .                  .                  I  .
 9.000D-09  5.000D-01 .       I0       .                  .                  I  .
 1.000D-08  5.000D-01 .       I0       .                  .                  I  .
 1.100D-08  5.000D-01 .       I0       .                .I                   $  .
 1.200D-08  5.000D-01 .       I0       .            I    .                   $  .
 1.300D-08  5.000D-01 .       I0       .         I       .                   $  .
 1.400D-08  5.000D-01 .       I0     I .                 .                   $  .
 1.500D-08  5.000D-01 .       I0       .                 .                   $  .
 1.600D-08  5.000D-01 .       I0    I  .                 .                   $  .
 1.700D-08  5.000D-01 .       I0      I.                 .                   $  .
 1.800D-08  5.000D-01 .       I0       .         I       .                   $  .
 1.900D-08  5.000D-01 .       I0       .             I   .                   I  .
 2.000D-08  5.000D-01 .       I0       .                 .                   I  .
 2.100D-08  5.000D-01 .       I 0      .                 .                   I  .
 2.200D-08  -5.000D-01.       I    0   .                 .                   I  .
 2.300D-08  5.000D-01 .       I         0                .                   I  .
 2.400D-08  5.000D-01 .       I       .             0    .                 I+ .
 2.500D-08  5.000D-01 .       I       .                  .0  +              +  .
 2.600D-08  5.000D-01 .       I       .             0    .                 0I .
 2.700D-08  5.000D-01 .       I   I    .                 .                   I  .
 2.800D-08  5.000D-01 .      I I      .                  .                   I  .
 2.900D-08  5.000D-01 .      I$       .                  .                   I  .
 3.000D-08  5.000D-01 .      I$       .                  .                   I  .
 3.100D-08  5.000D-01 .      I$       .                  .                   I  .
 3.200D-08  5.000D-01 .      I$       .                  .                   I  .
 3.300D-08  5.000D-01 .      I$       .                  .                   I  .
 3.400D-08  5.000D-01 .      I$       .                  .                   I  .
 3.500D-08  5.000D-01 .      I$       .                  .                   I  .
 3.600D-08  5.000D-01 .      I$       .                  .                   I  .
 3.700D-08  5.000D-01 .      I$       .                  .                   I  .
 3.800D-08  5.000D-01 .      I$       .                  .                   I  .
 3.900D-08  5.000D-01 .      I$       .                  .                   I  .
 4.000D-08  5.000D-01 .      I$       .                  .                   I  .
 4.100D-08  5.000D-01 .      I$       .                  .      +            0  .
 4.200D-08  5.000D-01 .      I$       .                  .                   A  .
 4.300D-08  5.000D-01 .      I$       .            +     .                   0  .
 4.400D-08  5.000D-01 .      I$     +  .                 .                   0  .
 4.500D-08  5.000D-01 .      I$       .                  .                   0  .
 4.600D-08  5.000D-01 .      I$       .                  .                   0  .
 4.700D-08  5.000D-01 .      I$       .                  .                   0  .
 4.800D-08  5.000D-01 .      I$       .                  .                   0  .
 4.900D-08  5.000D-01 .      I$       .                  .                   0  .
 5.000D-08  5.000D-01 .      I $      .                  .                   0  .
 5.100D-08  5.000D-01 .      I $      .                  .                   I  .
 5.200D-08  5.000D-01 .      I $      .                  .                   I  .
 5.300D-08  5.000D-01 .      I $      .                  .                   I  .
                     .-------------------------------------------------------------------------------------------

Y
0
       JOB CONCLUDED
0      TOTAL JOB TIME      49.66
```

Fig. 9.6   SPICE output for Fig. 9.4

```
•TWO BIT JOHNSON RING COUNTER •
•••••••••••••••••••••••••••••••••••••••••••••••

TABLE II-TOTAL POWER VERSUS TIME
WAVEFORM OF TOTAL POWER DISSIPATED IN CKT. BLOCK                      ------MOSIMR------

•••••••••••••••••••••••••••••••••••••••••••••••

TIME(N-SEC) TOTAL POWER                         TOTAL POWER(U-WATTS)----->

        0 000E+00      0 140E+04      0 280E+04      0 420E+04      0 560E+04      0 700E+04
  0 000E+00  0 497E+04                                                   T
  0 100E+01  0 497E+04                                                   T
  0 200E+01  0 496E+04                                                   T
  0 300E+01  0 494E+04                                                   T
  0 400E+01  0 483E+04                                                  T
  0 500E+01  0 452E+04                                            T
  0 600E+01  0 442E+04                                      T
  0 700E+01  0 441E+04                                      T
  0 800E+01  0 441E+04                                      T
  0 900E+01  0 442E+04                                      T
  0 100E+02  0 442E+04                                      T
  0 110E+02  0 442E+04                                      T
  0 120E+02  0 442E+04                                      T
  0 130E+02  0 442E+04                                      T
  0 140E+02  0 442E+04                                      T
  0 150E+02  0 442E+04                                      T
  0 160E+02  0 442E+04                                      T
  0 170E+02  0 445E+04                                       T
  0 180E+02  0 451E+04                                        T
  0 190E+02  0 462E+04                                          T
  0 200E+02  0 498E+04                                                T
  0 210E+02  0 531E+04                                                     T
  0 220E+02  0 548E+04                                                       T
  0 230E+02  0 549E+04                                                       T
  0 240E+02  0 520E+04                                                   T
  0 250E+02  0 515E+04                                                  T
  0 260E+02  0 519E+04                                                   T
  0 270E+02  0 501E+04                                                 T
  0 280E+02  0 498E+04                                                T
  0 290E+02  0 497E+04                                                T
  0 300E+02  0 497E+04                                                T
  0 310E+02  0 497E+04                                                T
  0 320E+02  0 497E+04                                                T
  0 330E+02  0 497E+04                                                T
  0 340E+02  0 497E+04                                                T
  0 350E+02  0 497E+04                                                T
  0 360E+02  0 497E+04                                                T
  0 370E+02  0 497E+04                                                T
  0 380E+02  0 497E+04                                                T
  0 390E+02  0 497E+04                                                T
  0 400E+02  0 497E+04                                                T
  0 410E+02  0 497E+04                                                T
  0 420E+02  0 497E+01                                                T
  0 430E+02  0 497E+04                                                T
  0 440E+02  0 497E+04                                                T
  0 450E+02  0 497E+04                                                T
  0 460E+02  0 497E+04                                                T
  0 470E+02  0 500E+04                                                T
  0 480E+02  0 507E+04                                                 T
  0 490E+02  0 517E+04                                                  T
  0 500E+02  0 553E+04                                                     T
  0 510E+02  0 586E+04                                                       T
  0 520E+02  0 604E+04                                                         T
  0 530E+02  0 596E+04                                                        T

JOB CONCLUDED                                              TOTAL JOB TIME   3 20
```

Fig. 9.7   MOSIMR power dissipation output for Fig. 9.4

lated. Let waveforms be computed in the order nodes are numbered: When computing $v_i^{(1)}(t)$ we use the waveforms, $v_1^{(1)}(t)$, $v_2^{(1)}(t)$ . . . $v_{i-1}^{(1)}(t)$, $v_{i+1}^{(0)}(t)$ . . . $v_m^{(0)}(t)$. For G-J, we would use $v_1^{(0)}(t)$, $v_2^{(0)}(t)$, $v_{i-1}^{(0)}(t)$, $v_{i+1}^{(0)}(t)$ . . . $v_m^{(0)}(t)$. Many iterations would be needed. One cannot stop with one iteration as in timing simulators. However, waveform relaxation can be applied to any circuit, digital or analog. Floating capacitors, pass transistors, feedback paths, etc., pose no problems.

Before actually applying the relaxation process, it is convenient to partition the network [11]. For example, the one-bit dynamic shift register cell shown in Fig. 9.8(a) and (b) is partitioned as shown in Fig. 9.8(c). Partition S_1 is used to compute $v_1(t)$. Other voltages which can influence v_1 are included in the circuit segment S_1 used to find $v_1(t)$. $v_1^{(k)}(t)$ represents the kth iteration for $v_1(t)$. A Seidel process has been assumed in the partitioning. When calculating $v_1^{(k)}(t)$, $v_2^{(k-1)}(t)$ can be treated as an independent source like $u_1(t)$ and $u_2(t)$. Actual transient analysis of the circuit segments can be done using some technique like Backward Euler. Associated models can be used to

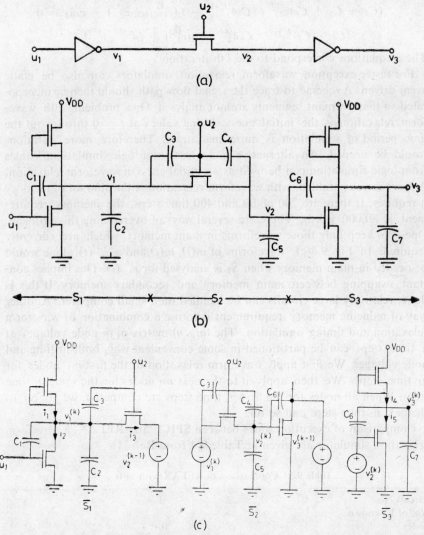

Fig. 9.8   (a) Logic representation of dynamic shift register cell. (b) Circuit representation of (a). (c) Partitioned representation of (b) (from [5])

represent the capacitors. In computing waveform $v_2^{(k)}(t)$, $v_1^{(k)}(t)$ and $v_3^{(k-1)}(t)$ are treated as independent sources. Similarly, in computing $v_3^{(k)}(t)$, $v_2^{(k)}(t)$ is treated as an independent source. While we have implied that all the unknowns are node voltages, this is not necessary. Lelarasmee et. al. [11] have given a partition scheme for Fig. 9.8 where the pass transistor current is also an unknown. This is more convenient when pass transistors are present. Referring to Fig. 9.8(c) again, the relevant decoupled differential equations corresponding to $\bar{S}_1$, $\bar{S}_2$ and $\bar{S}_3$ are

$$(C_1 + C_2 + C_3)\dot{v}_1^{(k)} - i_1(v_1^{(k)}) + i_2(v_1^{(k)}, u_1)$$
$$+ \, i_3(v_1^{(k)}, u_2, v_2^{(k-1)}) - C_1\dot{u}_1 - C_3\dot{u}_2 = 0$$

$$(C_4 + C_5 + C_6)\dot{v}_2^{(k)} - C_6\dot{v}_3^{(k-1)} - i_3(v_1^{(k)}, u_2, v_2^{(k)}) - c_4\dot{u}_2 = 0$$

$$(C_6 + C_7)\dot{v}_3^{(k)} - C_6\dot{v}_2^{(k)} - i_4(v_3^{(k)}) + i_5(v_3^{(k)}, v_2^{(k)}) = 0$$

These equations correspond to the kth iteration.

For faster execution waveform relaxation simulators can also be made event driven. A scheme to trace the signal flow path should then be incorporated so that dormant segments are not analysed. One problem with waveform relaxation is the initial guess. Using values at $t = 0$ throughout the time period of simulation is quite inaccurate. Therefore, more iterations would be needed. An alternative is to first do a logic simulation. Values from logic simulation can be used as an initial guess in waveform relaxation. A more serious problem with waveform relaxation is the amount of memory it requires. If there are 200 nodes and 400 time steps, the memory requirement is 80,000 words. There are several ways of overcoming this problem. One is to keep only those waveforms in main memory which are currently required. In Fig. 9.8(c), waveforms of $u_1(t)$, $u_2(t)$ and $v_2^{(k-1)}(t)$ alone would be needed in main memory when $\bar{S}_1$ is analysed for $v_1^{(k)}(t)$. This implies constant swapping between main memory and secondary memory. If this is done, relatively large circuits can be handled on a small computer. Another way of reducing memory requirement is to use a combination of waveform relaxation and timing simulation. The $(m \times n)$ matrix of m node voltages at n time steps can be partitioned in some convenient way, both in time and node voltages. We first apply waveform relaxation to the first m_1 nodes for n_1 time steps. We then apply it to the next m_2 nodes for the same n_1 time steps. When all nodes for the first n_1 time steps are computed, we go on to the next n_2 time steps and so on.

Comparison of execution times between SPICE and RELAX, a waveform relaxation simulator are given in Table 9.4 from Ref. [1].

Table 9.3   Comparison of RELAX with SPICE

Circuit No.	1	2	3	4	5
No. of Unknown nodes	4	8	16	27	45
No. of MOS DEVICES	6	21	42	131	263
CPU-SPICE (sec)	21.3	121.57	211.53	818.00	1334.80
CPU-RELAX (sec)	1.08	4.38	5.85	18.42	22.3
No. of RELAX Iterations	5	5	7	5	4
CPU SPICE/CPU RELAX	19.7	27.73	36.16	44.42	59.86

9.6 Mixed Mode Simulators

The kind of simulator used would depend on the application. For a well understood digital circuit, where timings are not critical, a simulation at the logic level may be sufficient. Where some timing information, though not

very accurate, is necessary one may use timing simulators. Timing simulators usually have simple device models. Capacitors permitted are linear and
often grounded. A digital circuit may have some critical components. An
example is the sense amplifier of a dynamic RAM. Such circuits may be
simulated using SPICE–like circuit simulators.

An alternative to having separate simulators is to have a mixed-mode
simulator [12]. One simulator contains circuit, timing and logic simulators.
We can analyse different parts of a circuit using different simulators.
Newton et. al. [12] give the simulation of a dynamic RAM as an example.
The row and column decoders are analysed using logic simulation. The
storage arrays and I/O control segments are analysed using timing simulation. Full-fledged circuit simulation is used to analyse the sense amplifiers.
An interface has to be built-in to convert the logic simulator voltage levels
to the circuit/timing simulator levels and vice versa. This is because the
logic simulator does not deal with actual voltages. A threshold element is
used to convert voltages to logic levels as shown in Fig. 9.9. Voltages below
V_1 are '0' and voltages above V_2 are '1'. Values between V_1' and V_2 correspond
to an unknown state 'u'. A logic to voltage (LTV) converter is used to convert
logic levels to voltages. Four-valued logic with '0', '1', 'u' and a high impedance state 'H' is assumed here [12]. Transitions from 0 to 1 or 1 to 0 are
made at a fixed rise time and fall time. During the high impedance state H
the previous value of voltage is maintained. During the 'u' state, the output

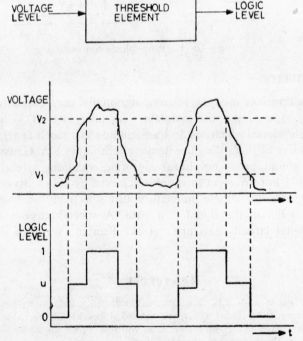

Fig. 9.9   Threshold element to convert voltages to logic levels

falls if it is high or rises if it is low. A LTV converter and associated waveforms are shown in Fig. 9.10.

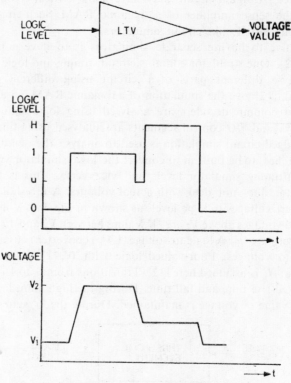

Fig. 9.10   Logic to voltage converter.

9.7 Summary

Non-linear transient analysis is computationally very expensive and time consuming. Special purpose simulators for analysing MOS digital circuits have been developed which can do transient analysis much faster than circuit simulators like SPICE. They use iterative techniques like Gauss-Newton to solve the simultaneous non-linear algebraic equations at each time step. Considerable improvement in speed is achieved by doing just one iteration. Like logic simulators, these simulators also selectively analyse only those gates which lie on the signal flow path. A special purpose simulator for NMOS digital circuits developed at IIT Kanpur, was described in this chapter.

References

1. A.R. Newton and A.L. Sangiovanni-Vincentelli, Relaxation-Based Electrical Simulation, IEEE Trans. CAD, vol. CAD-3, 4, pp. 308-331, Oct. 1984.
2. J.T. Deutsch, Circuit Simulator and Multiple Processor Spice up IC Design, Electronic Design, pp. 119-124, Dec. 1985.

3. B.R. Chawla, H.K. Gummel and P. Kozak, MOTIS—an MOS Timing Simulator, IEEE Trans. Ccts and Syst., vol. CAS–22, pp. 901–909, Dec. 1975.
4. S.P. Fan, M.Y. Hsueh, A.R. Newton and D.O. Pederson, MOTIS–C, A New Circuit Simulator for MOS LSI Circuits, Proc. IEEE Int. Symp. Ccts. and Syst. April, 1977.
5. E. Lelarasnee, A.E. Ruehli and A.L. Sangiovanni-Vencentelli, The Waveform Relaxation Method for Time Domain Analysis of Large Scale Integrated Circuits, IEEE Trans. CAD, vol. CAD–1, pp. 131–145. July, 1982.
6. J.R. Rice, Numerical Methods, Software and Analysis, Chap. 6, McGraw-Hill International Book Co., 1983.
7. J.R. Westlake, A Handbook of Numerical Matrix Inversion and Solution of Linear Equations, John Wiley and Sons, 1968.
8. J.M. Ortega and W.C. Rheinboldt, Iterative Solution of Non-linear Equations in Several Variables, Chap. 7, Academic Press, 1970.
9. S. Banerjee and R. Raghuram, MOSIMR: A Timing Simulator using Signal Propagation to Exploit Latency, First International Workshop on VLSI Design, Madras, India, Dec. 26–28, 1985.
10. H. Schichman and D.A. Hodges, Modelling and Simulation of IGFET Switching Circuits, IEEE J. Solid State Ccts., vol. SC-3, pp. 285–289, Sept. 1968.
11. A.R. Newton, Techniques for the Simulation of Large Scale Integrated Circuits, IEEE Trans. Ccts and Syst., vol. pp. 741–749, Sept. 1979.
12. R. Olson and S. McGrogan, Parallelizing SPICE in a Timesharing Environment, Applied Mathematics and Computation, pp. 185–195, vol. 20, Sept. 1986.

Problems

1. Show that the Gauss-Jacobi iteration method of solving a set of simultaneous linear equations may be written as

$$[x]^{(k-1)} = [x]^{(k)} + [C][r]^{(k)} \text{ where}$$

$$[r]^{(k)} = [B] - [A][x]^{(k)} \text{ and}$$

   $[C]$ is diagonal and given by

$$
\begin{bmatrix}
1/a_{11} & & & \\
& 1/a_{22} & & \\
& & \ddots & \\
& & & 1/a_{nn}
\end{bmatrix}
$$

2. Show that the Gauss-Seidel iterative method for linear equations converges in one step if $[A]$ is lower triangular.

3. Under what conditions would

   (a) Gauss–Jacobi converge in one step for linear equations.
   (b) Gauss-Jacobi and Gauss-Seidel proceed identically for linear equations.

4. Solve the set of simultaneous non-linear equations $(x_1 + 2x_2^2 = 1;\ x_2 + 2x_1^2 = 1)$ by the following techniques for any one of its four solutions.

   (a) Newton-Raphson
   (b) Jacobi-Newton with both loops carried to convergence
   (c) Jacobi-Newton with outer Jacobi loops carried to convergence and just one inner Newton iteration.
   (d) Repeat (b) and (c) for Gauss–Newton and Newton Gauss techniques.

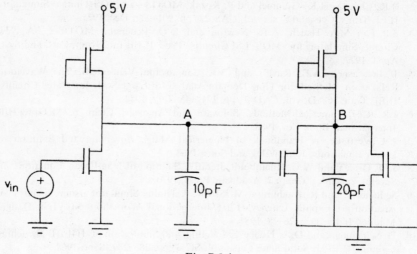

Fig. P.9.4

5. The circuit shown in Fig. P 9.4 is in steady state with $v_{in} = 5$ V. Assume $\beta_D = 10^{-4}$ A/V², $V_{TD} = 2$ V, $\beta_L = 2 \times 10^{-5}$ A/V², $V_{TL} = -2$ V..

(a) Find v_A and v_B

(b) At $t = 0$, v_{in} is switched instantaneously to OV. Write the node equations at A and B to get equations of the form

$$\frac{C dv_{A,B}}{dt} = f(v_A, v_B)$$

(c) Solve for $v_A(t)$, and $v_B(t)$ decoupling the equations at differential equation level and using G-S technique i.e. solve $v_A(t)$, then $v_B(t)$, then again $v_A(t)$ and so on. Use the Backward–Euler integration scheme.

(d) Replace the capacitors as in (c) by the associate model corresponding to the Backward–Euler scheme. At each time step solve for v_A and v_B doing just one combined Seidel–Newton iteration. What time step gives accurate results ?

(e) See that the Seidel–Newton technique with the Newton loop carried to convergence and with just one Seidel iteration gives accurate results for this circuit.

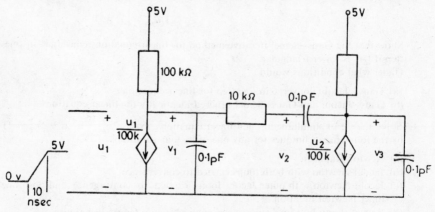

Fig. P.9.6

6. Introduce floating capacitors between gate and drain in the transistors of problem 5. Notice that

   (a) If a single Jacobi iteration in Jacobi–Newton is used, floating capacitors behave in exactly the same way as grounded capacitors.
   (b) For a single Seidel iteration of Seidel–Newton, floating capacitors connected to voltage nodes already calculated are properly accounted for, but the rest appear grounded capacitors.

7. (a) For the linear circuit shown in Fig. P 9.6, write the node equations in the form

$$[C][v] = [f(u_1, [v])]$$

   Here $[C]$ is the capacitance matrix and $[v]$ is the vector of node voltages.

   (b) $u_1(t)$ changes as indicated. Use the one sweep Gauss–Seidel technique to find $[v]$ at $t = \Delta t$. Use the Backward–Euler scheme to get a DC network (linear in this case). Take $\Delta t = 1$ ns.

   (c) Comment on the accuracy of the computation as applied to the given circuit.

10

Sensitivity and Optimisation

Circuit analysis procedure described upto this point have all assumed that component values are exact. However, it is well known that any component, either discrete or realised in an IC, has finite tolerance associated with its value. We talk of 5% and 10% tolerance in resistors. The β may easily vary over a factor of 2 from one transistor to another with the same number. Even if component values are known exactly, their variation with temperature could cause changes in the behaviour of the circuit. One very often would like to be able to predict how a certain response of a circuit would be affected by a change in some parameter of the circuit. The response could mean a DC voltage, small signal voltage gain, ratio of any two voltages/currents, etc. The parameter could be the value of any resistor, capacitor, inductance or controlled source. It could also be any parameter that describes another element like β of a transistor, temperature, reverse saturation current of a diode, etc. We would like to see how sensitive a certain response is to some parameter. If we are interested infinitesimal changes in the parameter, we should do a differential sensitive analysis. If the DC voltage V_0 is the response and p the parameter, the differential sensitivity is $\dfrac{\partial V_0}{\partial p}$.

Let there be m parameters $(p_1, p_2, \ldots p_m)$. Then the change in some response y at some value y_0 is given by the Taylor series expansion around y_0 as

$$y - y_0 = \Delta y = \frac{\partial y}{\partial p_1} dp_1 + \frac{\partial y}{\partial p_2} dp_2 \ldots \frac{\partial y}{\partial p_m} dp_m$$

$$+ \frac{1}{2!}\left(\frac{\partial^2 y}{\partial p_1 \partial p_1} dp_1 dp_1 + \frac{\partial^2 y}{\partial p_1 \partial p_2} dp_1 dp_2 \ldots \frac{\partial^2 y}{\partial p_1 \partial p_m} dp_1 dp_m \right.$$

$$+ \frac{\partial^2 y}{\partial p_2 \partial p_1} dp_2 dp_1 + \frac{\partial^2 y}{\partial p_2 \partial p_2} dp_2 dp_2 \ldots \frac{\partial^2 y}{\partial p_2 \partial p_m} dp_2 dp_m$$

$$+ \ldots \frac{\partial^2 y}{\partial p_m \partial p_1} dp_m dp_1 \ldots \left. \frac{\partial^2 y}{\partial p_m \partial p_m} dp_m dp_m \right) + \ldots$$

$$y = \sum_{i=1}^{m} \frac{\partial y}{\partial p_i} dp_i + \sum_{i=1}^{m} \sum_{j=1}^{m} \left(\frac{\partial^2 y}{\partial p_i \partial p_j} \right) dp_i dp_j + \ldots$$

For differential sensitivities the second and higher order partial derivatives are left out. In this chapter we will deal only with differential sensitivities.

Sometimes a normalised sensitivity is defined as

$$S_{norm} = \left(\frac{\partial y}{\partial p}\right)\bigg/\left(\frac{y_0}{p_0}\right) = \frac{\partial(\ln y)}{\partial(\ln p)}$$

This chapter also deals with optimisation. Though not directly related to sensitivity analysis, similar derivatives are needed. An example of optimisation is to get the transient response to an input step as close as possible to the desired response. The desired response may not be a step. If $r_d(t)$ is the desired response and $r(t)$ the actual response, we would like to minimise the following integral.

$$\int_0^T (r_d(t) - r(t))^2 \, dt$$

This may involve choosing proper values for many parameters in the circuit. The optimization in circuits is not different from the general optimisation problem and the same techniques are applicable for circuit optimisation.

10.1 Various Approaches to Sensitivity Analysis

If it is necessary to find the sensitivity of response y (like a voltage output) to some parameter p, the most straightforward way is to find the response y_0 to p_0. Then, for the same circuit we increment p to $p_0 + \Delta p$ and find $y = y_0 + \Delta y$. Then $\Delta y/\Delta p$ gives the required sensitivity. However, there are a number of problems with this approach.

1. Each time any sensitivity is needed two analyses of the circuit need to be done. If there are n responses and m parameters, a total of $2nm$ analyses are needed. This is a lot of work, especially for non-linear networks.

2. The operation we are doing is similar to numerical differentiation. We would be taking the difference of two large and almost equal quantities. The accuracy of the result would naturally be far less than that of either quantity.

A second approach is to use node voltage formulation or any other formulation. The actual equations to be solved are represented in terms of the solution of a chosen network with suitable inputs. The approach is described in detail in the next section. A third approach is to use something called the adjoint network. By analysing the actual network and the adjoint network any sensitivity can be found out. This approach is described in Section 10.3.

Whatever be the approach, the aim is to find the sensitivity matrix $[S]$ given below:

$$[S] = \begin{bmatrix} \dfrac{\partial y_1}{\partial p_1} & \dfrac{\partial y_1}{\partial p_2} & \cdots & \dfrac{\partial y_1}{\partial p_m} \\[2mm] \dfrac{\partial y_2}{\partial p_1} & \dfrac{\partial y_2}{\partial p_2} & \cdots & \dfrac{\partial y_2}{\partial p_m} \\[2mm] \vdots & & & \\[2mm] \dfrac{\partial y_n}{\partial p_1} & \dfrac{\partial y_n}{\partial p_2} & \cdots & \dfrac{\partial y_n}{\partial p_m} \end{bmatrix} \tag{10.1}$$

Each row gives the sensitivity of a single response to all the parameters and is called the row sensitivity. Each column gives the sensitivity of many responses to a gingle parameter and is called the column sensitivity. When we talk of sensitivity, we usually refer to DC networks (linear or non-linear) or to linear AC (sinusoidal steady state) networks. The sensitivity is calculated at a certain operating point. For example, the variation of a DC voltage in a transistor circuit with respect to β is calculated at some value β_0 and at some quiescent operating point. Sensitivities of AC networks are found at a particular frequency. It is convenient to define the phase and magnitude of a response as two different responses for AC circuits. The complex equations from any formulation (like node analysis) can be represented in AC circuits by twice the number of real equations.

Yet another approach, convenient when done with pencil and paper, is to actually differentiate the relevant expressions. We differentiate the function $y_i(p_j)$ to find $\dfrac{\partial y_i}{\partial p_j}$. Doing this on a computer requires symbolic manipulation which is quite difficult. Special programs exist for such differentiation, but are more complex than programs to find sensitivities.

10.2 Node Analysis Approach to Calculate Sensitivities Using Sensitivity Models [2]

Let us assume that the network is such that it can be solved using node analysis. For a general non-linear DC network, we can write the node equations as

$$f_1(V_1, V_2, \ldots V_n) = 0$$

$$f_2(V_1, V_2, \ldots V_n) = 0$$

$$\vdots$$

$$f_n(V_1, V_2, \ldots V_n) = 0$$

For the special case of a linear network, the above equations are linear. They can then be written in the form $[A][V] = [B]$ by moving the terms corresponding to the independent sources to the right hand side. $[A]$ then represents the node admittance matrix. These equations apply to linear AC networks as well if the various node voltages are interpreted as complex phasors. We will assume that any response y can be written in terms of the node voltages $[V]$. That is not very restrictive as this is true of most responses one would be interested in. For example, the current through a resistor can be written in terms of the voltages at the nodes between which it is connected. In two port networks, responses like $[H]$ matrix, $[Z]$ matrix, $[Y]$ matrix, etc., can be found from node voltages. If we allow some parameter p to vary, then the above equations will include p and can be written as

$$f_1(V_1, V_2 \ldots V_n, p) = 0$$
$$f_2(V_1, V_2 \ldots V_n, p) = 0$$
$$\cdot$$
$$\cdot$$
$$\cdot$$
$$f_n(V_1, V_2 \ldots V_n, p) = 0$$

The parameter p could be the value of any linear resistor (or L or C for AC networks), controlled source, independent source or any quantity defining the i–v characteristics of a non-linear element. Note that the various V_i are also functions of p. Then, for the ith equation, we get

$$\frac{df_i}{dp} = 0 = \frac{\partial f_i}{\partial p} + \frac{\partial f_i}{\partial V_1}\frac{dV_1}{dp} + \frac{\partial f_i}{\partial V_2}\frac{dV_2}{dp} \cdots \frac{\partial f_i}{\partial V_n}\frac{dV_n}{dp} \qquad (10.2)$$

One then gets the system of equations

$$\begin{bmatrix} \dfrac{\partial f_1}{\partial V_1} & \dfrac{\partial f_1}{\partial V_2} & \cdots & \dfrac{\partial f_1}{\partial V_n} \\[2mm] \dfrac{\partial f_2}{\partial V_1} & \dfrac{\partial f_2}{\partial V_2} & \cdots & \dfrac{\partial f_2}{\partial V_n} \\[1mm] \cdot & & \cdot & \\ \cdot & & \cdot & \\ \cdot & & \cdot & \\[1mm] \dfrac{\partial f_n}{\partial V_1} & \dfrac{\partial f_n}{\partial V_2} & \cdots & \dfrac{\partial f_n}{\partial V_n} \end{bmatrix} \begin{bmatrix} \dfrac{dV_1}{dp} \\[2mm] \dfrac{dV_2}{dp} \\[1mm] \cdot \\ \cdot \\ \cdot \\[1mm] \dfrac{dV_n}{dp} \end{bmatrix} = \begin{bmatrix} -\dfrac{\partial f_1}{\partial p} \\[2mm] -\dfrac{\partial f_2}{\partial p} \\[1mm] \cdot \\ \cdot \\ \cdot \\[1mm] -\dfrac{\partial f_n}{\partial p} \end{bmatrix} \qquad (10.3)$$

All the partial derivatives are evaluated at the operating point at which the sensitivity is needed. The $(n \times n)$ matrix in Eq. (10.3) is just the Jacobian (see Chapter 4). The equations in Eq. (10.3) constitute a system of linear equations which can be solved as described in Chapter 3. The vector solved for is the column sensitivity vector $\left[\dfrac{dV_1}{dp}, \dfrac{dV_2}{dp} \cdots \dfrac{dV_n}{dp} \right]$. The Jacobian is obtained automatically when solving the non-linear network at the operating point. The last iteration Jacobian can be retained and solved for a new right hand side. As the LU factors are available, this is very straightforward. From one analysis, the complete column sensitivity vector is obtained.

10.2.1 Linear Networks

Consider the special case of a DC linear network. The Jacobian matrix in Eq. (10.3) is then just the node admittance matrix. Let the parameter be the value of a resistor R in the network. If R is connected between nodes i and j, and f_i and f_j alone will be functions of R. Further $\dfrac{-\partial f_i}{dR}$ would be equal to $\dfrac{1}{R^2}(V_i - V_j)$ and $\left(\dfrac{-\partial f_j}{dR}\right)$ would be equal to $\dfrac{1}{R^2}(V_j - V_i)$, where V_i and V_j would be evaluated at the operating point. The operating point would have been found from an earlier analysis of the network. Equation 10.3 amounts to solving the same linear network for a new right hand side vector $[B]$ which has only two non-zero elements. A circuit interpretation can be given to this solution. Let the resistor R be changed to $R + dR$. Correspondingly, let the current through R change from I to $I + dI$.

$$I = \frac{V_{ij}}{R}$$

$$dI = \frac{1}{R}dV_{ij} - \frac{V_{ij}}{R^2}dR \quad \text{(Note } V_{ij} = V_i - V_j\text{)}$$

When writing the node equations at nodes i and j the extra current dI must be accounted for. This is done as shown in Fig. 10.1. The change dV_k at some node k is due to the extra current source $\frac{V_{ij}}{R^2}dR$. It can be found by removing other independent sources in the network i.e. short circuiting voltage sources and open circuiting current sources. This is exactly what we concluded when we applied Eq. (10.3) to a linear network. For example, let dV_B/dR_1 be the sensitivity required from Fig. 10.2(a). We then replace it by the circuit of Fig. 10.2(b). The circuit of Fig. 10.2(b) is analysed to find dV_B. V_A and V_B are found from the original circuit in Fig. 10.2(a). Solving these networks, we get

$$V_A = I_0 \frac{R_2(R_1 + R_3)}{R_2 + R_1 + R_3} \qquad V_B = \frac{I_0 R_2 R_3}{R_1 + R_2 + R_3}$$

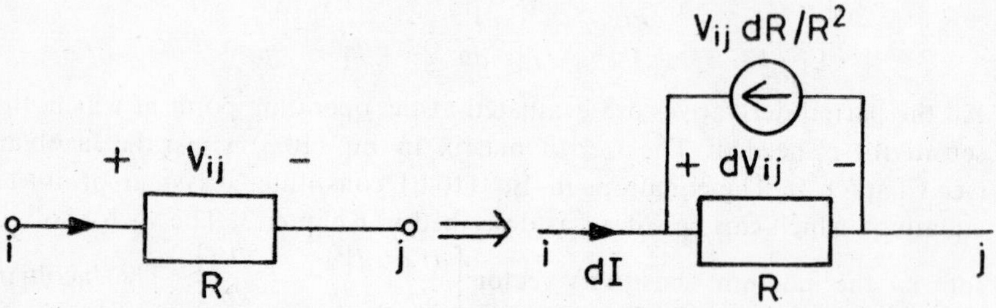

Fig. 10.1   Equivalent circuit to calculate effect of variation in some resistance R.

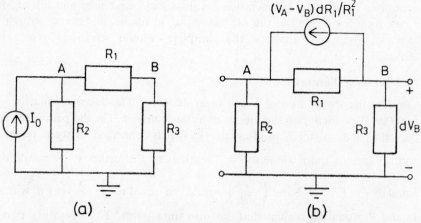

Fig. 10.2   Application of equivalent circuit of Fig. 10.1 to the circuit in (a) to get that in (b).

$$dV_B = -\frac{(V_A - V_B)}{R_1^2} \times \frac{R_1 R_3 dR_1}{R_1 + R_2 + R_3} = -\frac{(V_A - V_B)R_3 dR_1}{R_1(R_1 + R_2 + R_3)}$$

or,

$$\frac{dV_B}{dR_1} = \frac{-(V_A - V_B)R_3}{R_1(R_1 + R_2 + R_3)}$$

Direct differentiation of the expression for V_B gives the same result as can be easily verified.

It is not actually necessary to use a current source of value $V_{AB} dR_1/R_1^2$ in Fig. 10.2(b). This current source can be made equal to 1 A. Let the voltage at node B calculated from the circuit then be V_B'. Then

$$dV_B = \frac{V_B' V_{AB} dR_1}{R_1^2} \quad \text{or} \quad \frac{dV_B}{dR_1} = \frac{V_B' V_{AB}}{R_1^2}$$

The sensitivity of the voltage across any port with respect to any resistance is the product of the voltage across the resistance in the actual circuit and the voltage across the port in the sensitivity model divided by the square of the resistance. The circuit of Fig. 10.2(b) is the sensitivity model of the circuit in Fig. 10.2(a) if the current source in Fig. 10.2(b) is made equal to 1 A.

Sensitivity models can be constructed when other elements of the circuit vary. If a voltage source V_0 or a current source I_0 varies then the sources are replaced by dV_0 and dI_0 respectively in the sensitivity model. Other sources are removed. If a VCCS g_m varies we can write

$$dI = V_c dg_m + g_m dV_c$$

where V_c is the controlling branch voltage. Its sensitivity model equivalent is shown in Fig. 10.3. If the source $V_c dg_m$ is replaced by a 1A source and the response V_r' calculated, then

$$dV_r = V_r' V_c dg_m \quad \text{or} \quad \frac{dV_r}{dg_m} = V_r' V_c$$

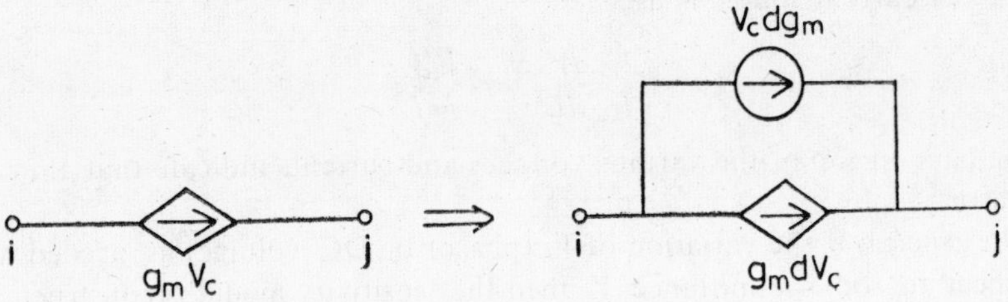

Fig. 10.3 Sensitivity model for a VCCS.

In other words, the sensitivity is the product of the response port voltage in the sensitivity model and the controlling voltage of the VCCS in the actual circuit.

For AC linear circuits capacitors and inductors may vary. Corresponding sensitivity models are shown in Fig. 10.4 and 10.5 respectively. For a capacitor

$$dI = j\omega C \, d\bar{V} + \bar{V} j\omega \, dC$$

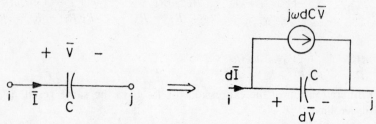

Fig. 10.4   Sensitivity model for a capacitor.

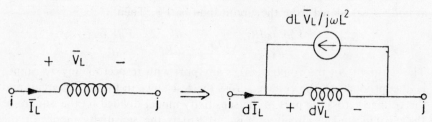

Fig. 10.5   Sensitivity model for an inductor.

If a 1 A current source is used in the sensitivity model and the response $\bar{V}'_r$ found, then

$$d\bar{V}_r = \bar{V}'_r \, \bar{V} j\omega \, dC$$

or

$$\frac{d\bar{V}_r}{dC} = \bar{V}'_r \, \bar{V} j\omega$$

Similarly for an inductor

$$d\bar{I}_L = \frac{d\bar{V}_L}{j\omega L} - \frac{\bar{V}_L dL}{j\omega L^2}$$

If a 1 A current source is used

$$\frac{d\bar{V}_r}{dL} = \frac{\bar{V}'_r \bar{V}_L}{j\omega L^2}$$

The bars on top of the various voltages and currents indicate that they are phasors.

In general, if the variation of V_r (phasor or DC voltage) is needed with respect to some admittance Y, then the sensitivity model is constructed as shown in Fig. 10.6. The sensitivity dV_r/dY is given by

$$\frac{dV_r}{dY} = V'_r \, V \tag{10.4}$$

where V'_r is the response in the sensitivity model and V the voltage across Y in the actual circuit. Note that the sensitivity model has the other independent sources removed. Once the actual circuit and the appropriate sensitivity model are analysed, any column sensitivity can be found.

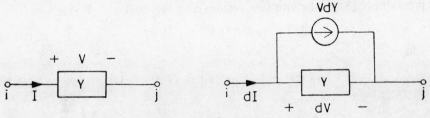

Fig. 10.6   Sensitivity model for a conductor.

So far we have assumed that just one parameter varies. If p stands for a vector of parameters $[p_1, p_2, \ldots p_m]$ then Eq. (10.2) can be rewritten as follows.

$$\frac{\partial f_1}{\partial p_j} + \frac{\partial f_1}{\partial V_1}\frac{\partial V_1}{\partial p_j} + \frac{\partial f_j}{\partial V_2}\frac{\partial V_2}{\partial p_j} + \cdots \frac{\partial f_i}{\partial V_n}\frac{\partial V_n}{\partial p_j} = 0$$

There are n functions and m parameters making a total of mn equations of this type. Equation (10.3) now takes the form given below. The total derivatives in Eq. (10.3) are replaced by partial derivatives.

$$
\begin{bmatrix}
\dfrac{\partial f_1}{\partial V_1} & \dfrac{\partial f_1}{\partial V_2} & \cdots & \dfrac{\partial f_1}{\partial V_n} \\[2ex]
\dfrac{\partial f_2}{\partial V_1} & \dfrac{\partial f_2}{\partial V_2} & \cdots & \dfrac{\partial f_2}{\partial V_n} \\[2ex]
\vdots & & & \\[2ex]
\dfrac{\partial f_n}{\partial V_1} & \dfrac{\partial f_n}{\partial V_2} & \cdots & \dfrac{\partial f_n}{\partial V_n}
\end{bmatrix}
\begin{bmatrix}
\dfrac{\partial V_1}{\partial p_1} & \dfrac{\partial V_1}{\partial p_2} & \cdots & \dfrac{\partial V_1}{\partial p_m} \\[2ex]
\dfrac{\partial V_2}{\partial p_1} & \dfrac{\partial V_2}{\partial p_2} & \cdots & \dfrac{\partial V_2}{\partial p_m} \\[2ex]
\vdots & & & \\[2ex]
\dfrac{\partial V_n}{\partial p_1} & \dfrac{\partial V_n}{\partial p_2} & \cdots & \dfrac{\partial V_n}{\partial p_m}
\end{bmatrix}
$$

$$(n \times n) \qquad\qquad (n \times m)$$

$$
=
\begin{bmatrix}
\dfrac{-\partial f_1}{\partial p_1} & \dfrac{-\partial f_1}{\partial p_2} & \cdots & \dfrac{-\partial f_1}{\partial p_m} \\[2ex]
\dfrac{-\partial f_2}{\partial p_1} & \dfrac{-\partial f_2}{\partial p_2} & \cdots & \dfrac{-\partial f_2}{\partial p_m} \\[2ex]
\vdots & & & \\[2ex]
\dfrac{-\partial f_n}{\partial p_1} & \dfrac{-\partial f_n}{\partial p_2} & \cdots & \dfrac{-\partial f_n}{\partial p_m}
\end{bmatrix}
\qquad (10.5)
$$

$$(n \times m)$$

In order to solve for the complete sensitivity matrix $[\partial V/\partial p]$, a system of linear equations must be solved m times, one column at a time. Looking at the circuit equivalent, m different sensitivity models must be constructed and solved-one for every parameter that varies. Branin [1] suggests a technique for avoiding this. Assume that the row sensitivity is needed i.e. the variation of one response V_r to m different parameters. Equation 10.5 is of the form $[J][S] = [B]$ and we want to solve for the first row of $[S]$. We premultiply $[S]$ by the unit row vector $[e_1] = [1, 0, 0, \ldots 0]$. Then

$$[e_1][S] = [e_1][J^{-1}][B] = [y_1^t][B] \tag{10.6}$$

where,

$$[e_1][J^{-1}] = [y_1^t] \text{ or } [y_1] = [J^{-1}]^t[e^1] \tag{10.7}$$

as

$$[J^t]^{-1} = [J^{-1}]^t$$

The vector $[y_1]$ can therefore be found by solving the system of equations $[J^t][y_1] = [e_1^t]$. Then $[e_1][S]$, the row sensitivity we need, is found from $[e_1][S] = [y_1^t][B]$. If the kth row of $[S]$ is needed we use the vector $[e_k] = [0, 0, \ldots 0, 1, 0, \ldots 0]$ instead of $[e_1]$. For a linear network $[J]$ is the node admittance matrix. The LU factors of J would be available. Then

$$[J^t] = [U^t][L^t]$$

The LU factors of $[J^t]$ needed to solve $[J^t][y_1] = [e_1^t]$ are also readily available. Using Branin's approach [1], the row sensitivity as well as column sensitivity can be found by solving two systems of linear equations. For column sensitivity the LU factors of both systems are the same. For row sensitivity they are transposed.

We were able to find column sensitivities by constructing the sensitivity model [2]. We would like to find a circuit representation of the approach described above to find row sensitivities. In other words, for linear networks, we should construct a circuit having a node admittance matrix which is the transpose of the node admittance matrix of the actual circuit. This is called the adjoint network model and is the subject of discussion in the next section. We may point out here that in the absence of controlled sources the node admittance matrix is symmetric. The adjoint network model has the same elements as the actual network, though the excitation is different. The excitation is the vector $[e_k^t]$. We pick up the discussion on adjoint networks in Section 10.3.

Example 10.1   For the linear network shown in Fig. 10.7(a) find

(a) the column sensitivity vector

$$\left[\frac{\partial V_A}{\partial R_2}, \frac{\partial V_B}{\partial R_2}, \frac{\partial V_C}{\partial R_2}\right]^t$$

(b) the row sensitivity vector

$$\left[\frac{\partial V_B}{\partial R_1}, \frac{\partial V_B}{\partial R_2}, \frac{\partial V_B}{\partial R_3}, \frac{\partial V_B}{\partial R_4}\right]$$

Solution: Writing the node equations at A, B and C

$$f_A = \frac{V_A - V_B}{R_1} + 2V_B - 1 = 0$$

$$f_B = \frac{V_B - V_A}{R_1} + \frac{V_B}{R_2} + \frac{V_B - V_C}{R_3} = 0$$

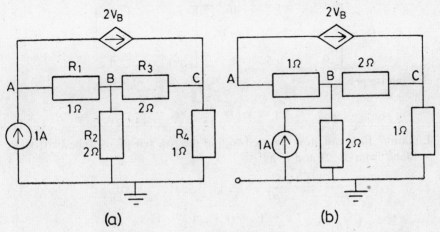

Fig. 10.7   (a) Circuit for which sensitivities are to be calculated. (b) Sensitivity model for (a).

$$f_C = \frac{V_C}{R_4} + \frac{V_C - V_B}{R_3} - 2V_B = 0$$

Substituting values for R_1, R_2, R_3, R_4 and rewriting in matrix form, we get

$$\begin{bmatrix} 1 & 1 & 0 \\ -1 & 2 & -0.5 \\ 0 & -2.5 & 1.5 \end{bmatrix} \times \begin{bmatrix} V_A \\ V_B \\ V_C \end{bmatrix} = \begin{bmatrix} 1 \\ 0 \\ 0 \end{bmatrix}$$

Solving, $\qquad V_A = \frac{7}{13}V, \; V_B = \frac{6}{13}V, \; V_C = \frac{10}{13}V$

(a) The sensitivity model is as shown in Fig. 10.7(b). Solving the network we get

$$V'_A = \frac{-6}{13}V, \; V'_B = \frac{6}{13}V, \; V'_C = \frac{10}{13}V$$

Therefore

$$\frac{\partial V_A}{\partial R_2} = \frac{V'_A V_B}{R_2^2} = \frac{-9}{169} \; V/\Omega$$

$$\frac{\partial V_B}{\partial R_2} = \frac{V'_B V_B}{R_2^2} = \frac{9}{169} \; V/\Omega$$

$$\frac{\partial V_C}{\partial R_2} = \frac{V_C V_B}{R_2^2} = \frac{15}{169} \; V/\Omega$$

(b) We need to find the second row of the sensitivity matrix. We use Eq. (10.7) to solve for the intermediate quantity $[y_2]$. The transpose of the node admittance matrix is now used. $[y_2]$ is obtained from

$$\begin{bmatrix} 1 & -1 & 0 \\ 1 & 2 & -2.5 \\ 0 & -0.5 & 1.5 \end{bmatrix} \begin{bmatrix} y_{21} \\ y_{22} \\ y_{23} \end{bmatrix} = \begin{bmatrix} 0 \\ 1 \\ 0 \end{bmatrix}$$

Solving we get

$$y_{21} = y_{22} = \frac{6}{13} \quad y_{23} = \frac{2}{13}$$

Equation 10.6 can now be used to find the row sensitivity. The matrix $[B]$ is found from f_A, f_B and f_C as

$$[B] = \begin{bmatrix} \dfrac{V_A - V_B}{R_1^2} & 0 & 0 & 0 \\[2ex] \dfrac{V_B - V_A}{R_1^2} & \dfrac{V_B}{R_2^2} & \dfrac{V_B - V_C}{R_3^2} & 0 \\[2ex] 0 & 0 & \dfrac{V_C - V_B}{R_3^2} & \dfrac{V_C}{R_4^2} \end{bmatrix}$$

$$= \begin{bmatrix} \dfrac{1}{13} & 0 & 0 & 0 \\[2ex] \dfrac{-1}{13} & \dfrac{3}{26} & \dfrac{-1}{13} & 0 \\[2ex] 0 & 0 & \dfrac{1}{13} & \dfrac{10}{13} \end{bmatrix}$$

$$\begin{bmatrix} \dfrac{\partial V_B}{\partial R_1}, & \dfrac{\partial V_B}{\partial R_2}, & \dfrac{\partial V_B}{\partial R_3}, & \dfrac{\partial V_B}{\partial R_4} \end{bmatrix}$$

$$= \begin{bmatrix} \dfrac{6}{13}, & \dfrac{6}{13}, & \dfrac{2}{13} \end{bmatrix} \begin{bmatrix} \dfrac{1}{13} & 0 & 0 & 0 \\[2ex] \dfrac{-1}{13} & \dfrac{3}{26} & \dfrac{-1}{13} & 0 \\[2ex] 0 & 0 & \dfrac{1}{13} & \dfrac{10}{13} \end{bmatrix}$$

$$= \begin{bmatrix} 0, & \dfrac{9}{169}, & \dfrac{-4}{169}, & \dfrac{20}{169} \end{bmatrix}$$

10.2.2 Non-linear Networks

The concepts developed in Section 10.2.1 are easily extended to non-linear circuits. If only one parameter is allowed to vary, then Eq. (10.3) can be used. The matrix $\left[\dfrac{\partial f}{\partial v}\right]$ represents the Jacobian. This would have been evaluated when the actual circuit was solved using Newton-Raphson. It represents the node admittance matrix of the linearised equivalent (see Chapter 4). The sensitivity model is this linear circuit with suitable sources added. Depending on which element the varying parameter describes, the location of the sources vary.

Let the parameter p belong to a non-linear resistor defined by $I = g(V, p)$. Then

$$dI = \frac{\partial q}{\partial p}\, dp + \frac{\partial q}{\partial V}\, dV$$

The representation of the non-linear resistor in the sensitivity model is as shown in Fig. 10.8. The change in some response dV_r due to a change in parameter p can be found, as before, by removing all other sources and replacing the non-linear resistor by the model shown in Fig. 10.8. If the current source $\frac{\partial q}{\partial p}\, dp$ is made equal to 1 A and the corresponding response is V_r' then

$$dV_t = V_r'\, \frac{\partial q}{\partial p}\, dp$$

or

$$\frac{dV_r}{dp} = V_r'\, \frac{\partial q}{\partial p} \tag{10.8}$$

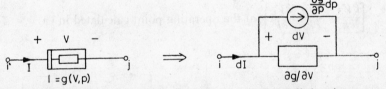

Fig. 10.8   Sensitivity model for non-linear voltage controlled resistor.

A complete derivative is used in the left hand side as p is the only parameter which varies. The rules for constructing the sensitivity model for a non-linear network are

1. Construct the linearised equivalent at the operating point.
2. Remove all sources i.e. replace voltage sources by short circuits and current sources by open circuits.
3. If the parameter p that varies belongs to a VCCS, resistor or independent current source, add a source of 1 A in parallel. For a voltage source add a 1 V source in series.
4. Find the response V_r' for the sensitivity model. Then

$$\frac{dV_r}{dp} = V_r'\, \frac{\partial q}{\partial p} \quad \text{for a resistor } I = g(V, p)$$

$$= V_r'\, \frac{\partial g_m}{\partial p} \quad \text{for a VCCS } I = g_m(V_c, p)$$

$$= V_r'\, \frac{\partial I_s}{\partial p} \quad \text{for an independent current source } I_s(p)$$

$$= V_r' \frac{\partial V_s}{\partial p} \text{ for an independent voltage source } V_s(p) \qquad (10.9)$$

These relations apply to linear resistors and VCCS' as well. For a linear resistor

$$I = g(V) = \frac{V}{R}$$

$$\frac{\partial q}{\partial R} = \frac{-V}{R^2}$$

$$\frac{dV_r}{dR} = \frac{-V_r'V}{R^2}$$

which is the same as the earlier relation. The negative sign is because the current source of 1 A in Fig. 10.8 is opposite in direction to that in Fig. 10.2. This relation for a linear resistor holds even if other elements in the actual network are non-linear.

Row sensitivities can again be found from Eqns (10.6) and (10.7).

Example 10.2: For the non-linear network shown in Fig. 10.9(a) find

(a) the operating point or quiescent solution for $k = 2$, $R_1 = R_2 = R_3 = R_4 = 1\,\Omega$

(b) the column sensitivity vector

$$\left[\frac{\partial V_A}{\partial k}, \frac{\partial V_B}{\partial k}, \frac{\partial V_C}{\partial k} \right]^t \text{ at the operating point calculated in (a).}$$

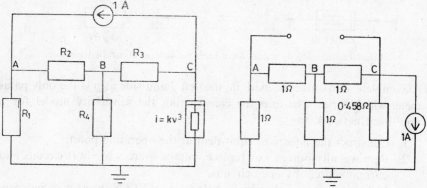

Fig. 10.9 (a) Non-linear circuit for which sensitivity is to be found.
(b) Sensitivity model.

Solution: (a) Writing node equations at A, B and C we get

$$V_A\left[\frac{1}{R_1} + \frac{1}{R_2}\right] - \frac{V_B}{R_2} = 1$$

$$V_B\left[\frac{1}{R_2} + \frac{1}{R_3} + \frac{1}{R_4}\right] - \frac{V_A}{R_2} - \frac{V_C}{R_3} = 0$$

$$\frac{V_C}{R_3} + kV_C^3 - \frac{V_B}{R_3} = -1$$

These equations, after substitution of values, can be solved using Newton-Raphson. Actually, they are simple enough that they can be solved directly. Solution yields

$$V_A = 0.48 \text{ V}, \quad V_B = -0.412 \text{ V}, \quad V_C = -0.603 \text{ V}$$

(b) In order to find the sensitivity model, the linearised equivalent of the non-linear resistor must be found. It has a i–v relation $I = g(V, k) = kV^3$

$$\frac{\partial I}{\partial V} = 3kV^2 = 2.182 \text{ mhos for } k = 2 \text{ and } V_C = -0.603 \text{ V}$$

As k is the parameter being varied, the excitation for the sensitivity model is the 1 A source in parallel with the resistance equal to $\frac{1}{2.182} = 0.458 \ \Omega$. The sensitivity model is shown in Fig. 10.9(b). Solving the sensitivity model we get

$$V'_A = -0.0719 \text{ V}, \quad V'_B = -0.1438 \text{ V}, \quad V'_C = -0.360 \text{ V}$$

From Eq. (10.8)

$$\frac{\partial V_r}{\partial p} = V'_r \frac{\partial q}{\partial p}$$

Here we need

$$\frac{\partial q}{\partial k} = V^3 /_{V=V_C}$$

Therefore,

$$\frac{\partial q}{\partial k} = -0.219.$$

We now get

$$\frac{\partial V_A}{\partial k} = V'_A \frac{\partial q}{\partial k} = 0.0157 \text{ V}^4/\text{A}$$

$$\frac{\partial V_B}{\partial k} = V'_B \frac{\partial q}{\partial k} = 0.0315 \text{ V}^4/\text{A}$$

$$\frac{\partial V_C}{\partial k} = V'_C \frac{\partial q}{\partial k} = 0.0786 \text{ V}^4/\text{A}$$

10.3 Adjoint Network Approach

The sensitivity model approach is convenient for calculating column sensitivities i.e. variation of many responses with respect to one parameter. It was pointed out in the last section that row sensitivities could be calculated using Eqns (10.6) and (10.7). It involved solving a network with a node admittance matrix which is the transpose of that of the original circuit. Often, the only computing tool available is a program to analyse circuits. If we can find the network which corresponds to the transposed matrix, then its analysis gives row sensitivities. In Eqns (10.6) and (10.7), $[J]$ represents the node admittance matrix for a linear network. For a non-linear network, it represents the node admittance matrix of the linearised equivalent at the quiescent or operating point. Our job is to find a linear network which has

$[J]^t$ as the node admittance matrix. This network is referred to as the adjoint network. The following rules can be used to construct the adjoint network. In the discussion below, the term 'actual circuit' refers to the original circuit for linear circuits and to the linearised equivalent for non-linear circuits.

1. Linear resistors in the actual circuit are left as they are. These are reciprocal elements and make symmetric contributions to the node admittance matrix.

2. For a VCCS, the controlling branch and the controlled branch are exchanged in the adjoint network as shown in Fig. 10.10(a). Other controlled sources are transformed as shown in Fig. 10.10(b), (c) and (d). These models could be used, if a hybrid technique, as apposed to node analysis, is used for analysis.

3. In solving for $[y_1]$ in Eq. (10.7), a unit vector $[e_1^i]$ was used as the excitation. This corresponds to a 1 A current source across the branch where the variation in response is needed. If the response is V_{ij}, the voltage between nodes i and j, we take the current source flowing as from as j to i. All sources in the original circuit are removed i.e. voltage sources shorted, current sources opened.

Once the adjoint network is constructed it can be analysed and all node voltages found. These voltages correspond to the vector $[y_1]$ (more generally $[y_i]$) in Eq. (10.7). We will now call this voltage vector $[V']$, with the prime standing for adjoint. We must next find $[B]$ and multiply $[V']$ and $[B]$ to get the row sensitivity. The matrix $[B]$, shown in Eq. (10.5), has elements of the form $-\partial f/\partial p$. Let p be a parameter of an element connected between nodes a and b. If the current from a to b is $g(p)$, then the column in $[B]$ corresponding to p has non-zero entries in row a and row b alone. These entries are $-\dfrac{\partial q}{\partial p}$ and $+\dfrac{\partial q}{\partial p}$ respectively. Therefore, the row sensitivity is given by $[y][B] = [V'][B] = -(V_a' - V_b')\dfrac{\partial q}{\partial p}$. For a linear resistor R connected between nodes a and b,

$$\frac{\partial q}{\partial p} = \frac{\partial q}{\partial R} = -\frac{V_{ab}}{R^2}$$

$$\frac{\partial V_r}{\partial R} = \frac{V_{ab}' V_{ab}}{R^2}$$

In general for a non-linear resistor or VCCS, connected between nodes a and b, of the form $I = g(V, p)$ (where V can be a controlling voltage), we have

$$\frac{\partial V_r}{\partial p} = -(V_a' - V_b')\frac{\partial q}{\partial p} = -V_{ab}'\frac{\partial q}{\partial p} \tag{10.10}$$

where V_{ab}' is the voltage between nodes a and b in the adjoint network.

We have so far assumed that our method of analysis is node analysis. This restricted the responses to node voltages and network elements to linear resistors, linear and non-linear VCCS and voltage controlled non-linear

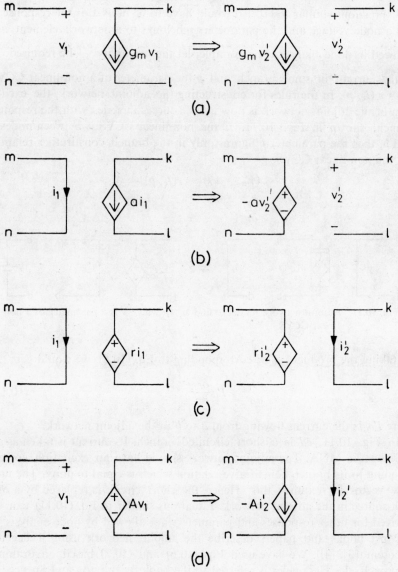

Fig. 10.10 Adjoint network representations for (a) VCCS (b) CCCS (c) CCVS (d) VCVS.

resistors. Excitation was due to independent current sources. However, adjoint networks can be constructed for any general network. Then hybrid technique of analysis would have to be used, for example, modified node analysis (MNA) could be used. Figure 10.10 shows how the adjoint network can be constructed for any linear controlled source. For a general network, the linearised equivalent and corresponding Jacobian can be constructed as described in Chapter 4. The Jacobian would correspond to the modified node admittance of the linearised equivalent if MNA is used.

Expressions similar to (10.10) would have to be derived if the response is not a node voltage and the parameters p belongs to a network element not allowed in node analysis. For example, let the sensitivity $\dfrac{\partial I_r}{\partial p}$ be required. I_r is the current in some branch and p the parameter of a non-linear CCVS $V = r(I_c, p)$. In the rules for constructing the adjoint network, the excitation in the adjoint network is now a 1 V sources in series with the response branch, shown in Fig. 10.11. If the non-linear CCVS is between nodes a and b, then the parameter p figures only in the branch constitutive relation for the non-linear CCVS.

$$f = (V_a - V_b) - r(I_c, p) = 0$$

$$-\frac{\partial f}{\partial p} = \frac{\partial r}{\partial p}$$

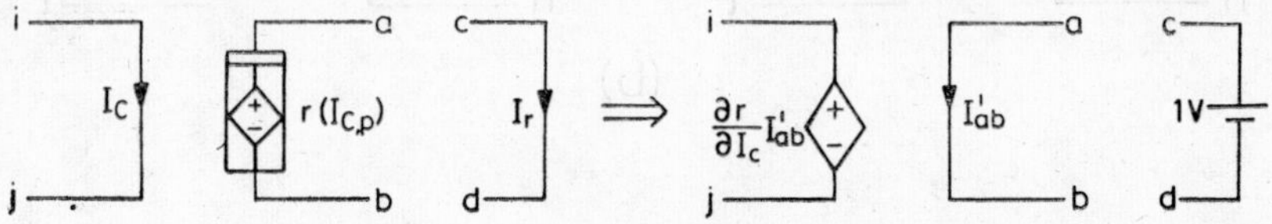

Fig. 10.11   Adjoint network model to find $\partial I_r / \partial p$ where p is a parameter of a non-linear CCVS.

In taking the product $[y][B]$ corresponding to Eq. (10.6), we would get

$$\frac{\partial I_r}{\partial p} = I'_{ab} \frac{\partial r}{\partial p} \tag{10.11}$$

Here I'_{ab} is the current flowing from a to b in the adjoint network.

In Fig. 10.11, cd is a short circuited branch. Its current is taken as an unknown in MNA. The adjoint network should have an excitation corresponding to its branch constitutive relation which is equal to unity. The unit row vector $[e]$ requires this. Hence the short circuit is replaced by a one volt source in the adjoint network. Equations similar to Eq. (10.11) can be derived for other responses and parameters not allowed by node analysis.

Most books and papers describe the adjoint network using Tellegen's theorem ([3], [4]). We have used Eqs (10.6) and (10.7) based on Branin's approach [1]. The reader is referred to the book by Brayton and Spence [2] for a detailed treatment. These authors also describe how noise analysis can be done using adjoint networks. Noise analysis is not discussed here.

Whether the sensitivity model or the adjoint network is used depends on whether row or column sensitivities are required. If the variation of many responses to one or two parameters is needed, sensitivity models would be preferred. If the variation of one or two voltages with many parameters is needed, the adjoint network may have some advantages.

Example 10.3:   For the non-linear circuit shown in Fig. 10.9(a) construct the adjoint network and find the row sensitivity vector $\left[\dfrac{\partial V_c}{\partial k}, \dfrac{\partial V_c}{\partial R_1}, \dfrac{\partial V_c}{\partial R_4}\right]$.

Assume $k = 2$, $R_1 = R_2 = R_3 = R_4 = 1$ to find the quiescent operating point.

Solution: The operating point solution was found in Example 10.2 to be

$$V_A = 0.48 \text{ V}, \ V_B = -0.412 \text{ V}, \ V_C = -0.603 \text{ V}$$

The resistance of the linearised equivalent of the non-linear resistor was found to be $0.458 \ \Omega$. The adjoint network, therefore, is as shown in Fig. 10.12. The 1 A current source is across the response port. It so happens that the sensitivity model of Fig. 10.9(b) is the same as the adjoint network model for this case, except that the current source direction is reversed. Solving the adjoint network, we get

$$V'_A = 0.0719 \text{ V}, \ V'_B = 0.1438 \text{ V}, \ V'_C = 0.360 \text{ V}$$

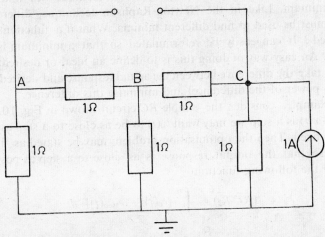

Fig. 10.12   Adjoint network corresponding to Fig. 10.9(a) where the change in the response V_C is required.

From Eq. (10.10)

$$\frac{\partial V_C}{\partial k} = -V'_C \frac{\partial q}{\partial k} \quad \text{where } g(V, k) = kV^3$$

$$= (-0.0360) \times (-0.219) = 0.0786 \text{ V}^4/\text{A}^2$$

This agrees with the result in Example 10.2. As R_1 and R_4 are linear resistors

$$\frac{\partial V_C}{\partial R_1} = \frac{V'_A V_A}{R_1^2} = \frac{0.0719 \times 0.480}{1^2} = 0.0345 \text{ A}$$

$$\frac{\partial V_C}{\partial R_4} = \frac{V'_B V_B}{R_4^2} = \frac{0.1438 \times (-0.0412)}{1^2} = 0.0059 \text{ A}$$

10.4 Optimisation Concepts

Whether a circuit designer is aware of it or not, he is usually doing optimisation. He starts with a given topology and chooses values for the components so that the performance is best or optimum. In addition, there are constraints to component values. Further, in optimising one performance requirement some other requirement may deteriorate. In such cases, the designer assigns weights to these different requirements. It is true that a designer, because of his experience, can juggle many parameters. Optimisation programs will be far less flexible. Still an optimisation program readily available can be an extra tool the designer uses to juggle his parameters. Where the process of design can be well formulated, optimisation programs may be very useful.

The first step is to express the requirement as a function of the many parameters which can be varied. We then attempt to find the absolute minimum of the function. Most optimisation methods will find only the local minimum. Like in the Newton-Raphson technique, different initial guesses must be used to find different minima. What if a function is to be maximised? It can easily be reformulated so that a minimum is what we look for. An easy way of doing this is to define an ideal or desired response. We then take the difference between the actual response and desired response (or same power of the difference) and minimise this difference.

For example, consider the simple RC circuit shown in Fig. 10.13 [5]. If the input $e(t)$ is a step we may want $v_0(t)$ to be as close to a step as possible. Let C be fixed. Then the optimisation problem may be stated as finding R_1 and R_2 so that the output response is as close to a step as possible. We minimise the following function

$$f(R_1, R_2) = \int_0^\infty [v_0(t) - v_{step}(t)]^2 \, dt$$

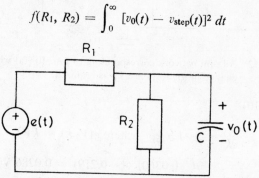

Fig. 10.13   Circuit for which rise time is to be optimised.

Very often a designer has to trade-off between conflicting requirements. Figure 10.14 shows a NMOS inverter. Let all parameters be fixed except the β (transconductance parameter) of the depletion mode load transistor. The quantity β is directly proportional to the ratio W/l for the transistor. It is well known that if β is large, the input-output characteristic is sharp, but the transient response is poor. For optimisation the designer must attach some weight to each of these performance characteristics viz., input-output DC characteristics and transient response. Let these weights be W_1

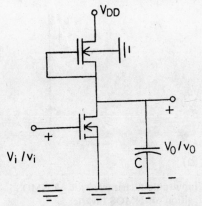

Fig. 10.14   NMOS inverter for which both pulse response and input-output characteristic are to be optimised.

and W_2. Then the function to be optimised can be written as

$$f(\beta) = \int_0^{V_{cc}} W_1^2(V_i)[V_0 - V_{d1}]^2 \, dV_i + \int_0^\infty W_2^2(t)[v_0 - v_{d2}]^2 \, dt$$

where,

V_{d1} = output corresponding to desired input-output DC characteristic

V_0 = actual DC output

V_i = DC input

V_{d2} = desired output response for an output low to high transition

v_0 = actual output response

W_1, W_2 = are the weights

While V_0 and V_{d1} are functions of V_i, v_0 and v_{d2} are functions of time t. W_1 and W_2 can either be constant or functions of V_i and t respectively. If W_1 is a function of V_i some particular part of the characteristic can be given more (or less) importance. Again, if W_2 is a function of t, some particular part of the waveform can be accentuated or downplayed. Figures 10.15(a) and (b) show V_0, V_{d1}, v_0, v_{d2}, etc. Let us for simplicity assume W_1 and W_2 to be constants. Then, depending on the values of W_1 and W_2, a different value of β will be obtained on optimisation. At the extremes, $\beta = \infty$ if $W_1 = 0$ and $\beta_1 = 0$ for $W_2 = 0$. If we plot I_1 as a function of I_2, we get a variation as shown in Fig. 10.16. Here

$$I_1 = \int_0^{V_{cc}} [V_0(V_i) - V_{d1}(V_i)]^2 \, dV_i$$

and

$$I_2 = \int_0^\infty [v_0(t) - v_{d2}(t)]^2 \, dt$$

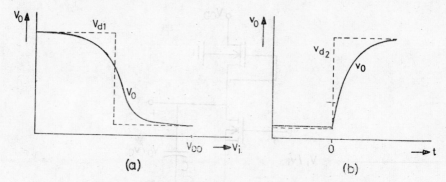

Fig. 10.15   (a) Input-output characteristic of NMOS inverter (b) ϕ to 1
transition of NMOS inverter

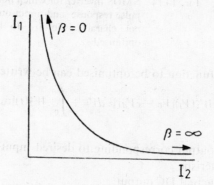

Fig. 10.16   Variation of optimised I_1 and
I_2 with variation in the weights
W_1 and W_2.

From the figure, the designer can decide where his optimum lies and accordingly choose β.

Another application of optimisation is in filters. If $V_0(\omega)$ is the frequency response obtained and $V_d(\omega)$ the desired response, then one may optimise the function

$$f([x]) = \int_0^\infty [V_0(\omega) - V_d(\omega)]^2 \, d\omega$$

where $[x]$ is the list of parameters to be varied.

The circuit optimisation problem is not different from the general optimisation problem. Once the function to be optimised is defined, one of the standard optimisation techniques can be applied. The choice of technique would, of course, depend on the function. The following section on optimisation techniques is only to introduce the reader to these techniques. For a detailed treatment of general optimisation techniques the reader is referred to the book by Rao [6]. References [2], [5] and [7] discuss optimisation as applied to circuits.

10.5 Optimisation Techniques

Let the function to be optimised be given by

$$f(x_1, x_2, \ldots x_n) = f([x])$$

Our aim is to find the vector $[x]$ such that $f([x])$ is minimum. Ideally we should find the global minimum. But the techniques discussed will only give the local minimum. Different starting values must be used to find the global minimum.

All common optimisation methods start with an initial guess for the vector $[x]$. We then keep moving according to some rules in a direction along which the function f decreases. There are three different operations usually done. These are

(a) Finding the direction along which one should move
(b) Finding how far one should move in the chosen direction
(c) Evaluating the gradient of f, ∇f, with respect to the various x_i i.e.,

finding $\left[\dfrac{\partial f}{\partial x_1}, \dfrac{\partial f}{\partial x_2}, \cdots \dfrac{\partial f}{\partial x_n}\right]$.

The most obvious choice for the direction is the one in which f decreases most rapidly. This direction is the negative gradient direction, $(-\nabla f)$. The method is called the method of steepest descent. If $[x^{(k)}]$ is the vector after k iterations then,

$$[x^{(k+1)}] = [x^{(k)}] - \alpha^{(k)} \frac{\nabla f^{(k)}}{|\nabla f^{(k)}|} = [x^{(k)}] + \alpha^{(k)}[S^{(k)}] \qquad (10.12)$$

where $[S^{(k)}]$ is a unit vector in the negative gradient direction and $\alpha^{(k)}$ is a scalar. The quantity $\alpha^{(k)}$ determines how far one should move in the negative gradient direction. We will shortly see how $\alpha^{(k)}$ may be determined.

The method of steepest descent does not converge very rapidly for elongated contours. Figure 10.17 shows two kinds of contours for a function of two variables. The contours represent lines of constant f. Figure 10.17(a) represents a function with a valley in the shape of an inverted cone with circular contours. Whatever be the initial guess the negative gradient points

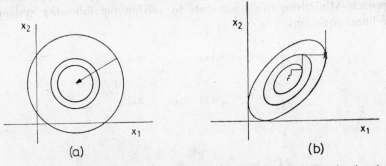

Fig. 10.17 Method of steepest descent converges (a) immediately for circular (cone shaped) contours and (b) very slowly for elongated contours.

to the direction of the minimum. If α is chosen appropriately convergence takes place in one step. Figure 10.17(a) shows an enlongated contour. In such cases convergence is very slow. A class of methods called Newton methods provide a better alternative. Information about the contour is used in finding the direction in which one should move. In these methods a quadratic approximation to the function $f([x])$ is made in the region of $[x^{(k)}]$. From the Taylor series expansion around $[x^{(k)}]$ we can write

$$f([x^{(k+1)}]) = f([x^{(k)}]) + \nabla f^{(k)}[\Delta x^{(k)}] + \tfrac{1}{2}[H^{(k)}][\Delta x^{(k)}][\nabla x^{(k)}] \qquad (10.13)$$

where $[H^{(k)}] = (n \times n)$ matrix called the Hessian given by

$$\begin{bmatrix} \dfrac{\partial^2 f}{\partial x_1^2} & \dfrac{\partial^2 f}{\partial x_1 \partial x_2} & \dfrac{\partial^2 f}{\partial x_1 \partial x_3} & \cdots & \dfrac{\partial^2 f}{\partial x_1 \partial x_n} \\[3mm] \dfrac{\partial f}{\partial x_2 \partial x_1} & \dfrac{\partial^2 f}{\partial x_2^2} & \dfrac{\partial^2 f}{\partial x_2 \partial x_3} & \cdots & \dfrac{\partial^2 f}{\partial x_2 \partial x_n} \\[3mm] \vdots & & & & \\[3mm] \dfrac{\partial^2 f}{\partial x_n \partial x_1} & \dfrac{\partial^2 f}{\partial x_n \partial x_2} & \dfrac{\partial^2 f}{\partial x_n \partial x_3} & \cdots & \dfrac{\partial^2 f}{\partial x_n^2} \end{bmatrix} / [x^{(k)}]$$

and
$$[\Delta x^{(k)}] = [x^{(k+1)}] - [x^{(k)}]$$

We find $[x^{(k+1)}]$ by trying to find out where $f([x^{(k+1)}])$ defined in Eq. (10.13) is minimum. Differentiating with respect to $[x^{(k+1)}]$ and equating to zero,

$$0 = \nabla f^{(k)} + [H^{(k)}][\Delta x^{(k)}]$$

or
$$[\Delta x^{(k)}] = - [H^{(k)}]^{-1} \nabla f^{(k)}$$

or
$$[x^{(k+1)}] = [x^{(k)}] - [H^{(k)}]^{-1} \nabla f^{(k)} \qquad (10.14)$$

This indicates that one should move in the direction indicated by $[H^{(k)}]^{-1} \nabla f^{(k)}$ in finding $[x^{(k+1)}]$ from $[x^{(k)}]$. Newton methods are an improvement over the method of steepest descent in the sense that an extra term of the Taylor expansion is used. Equation 10.14 can be arrived at from a different approach. Minimising $f([x])$ amounts to solving the following system of non-linear equations

$$\frac{\partial f}{\partial x_1} = 0 = g_1(x_1, x_2, \ldots, x_n)$$

$$\frac{\partial f}{\partial x_2} = 0 = g_2(x_1, x_2, \ldots, x_n)$$

$$\vdots$$

$$\frac{\partial f}{\partial x_n} = 0 = g_n(x_1, x_2, \ldots, x_n)$$

Each equation above is a non-linear equation in n unknowns $[x_{1,}, x_2, \ldots x_n.]$ These can be solved using the Newton-Raphson technique to give

$$[x^{(k+1)}] = [x^{(k)}] - [J^{(k)}]^{-1}[g([x^{(k)}])]$$

$$= [x^{(k)}] - [H^{(k)-1}\nabla f^{(k)}$$

as $$[g] = \nabla f \quad \text{and} \quad [J] = [H]$$

The above equation is the same as Eq (10.14). As an improvement over Eq. (10.14), it is usually rewritten as

$$[x^{(k+1)}] = [x^{(k)}] + \alpha^{(k)}[p^{(k)}] \tag{10.15}$$

where $[p^{(k)}]$ is a unit vector the same direction as the vector $(-[H^{(k)}]^{-1}\nabla f^{(k)})$.

Equation 10.15 is now in the same form as Eq. 10.12 with α to be determined. Newton methods are more difficult to implement as the Hessian and its inverse need to be found. Various quasi-Newton techniques have emerged which evaluate $[H]$ and $[H]^{-1}$ using different approximations. The reader is directed to references [2], [6] and [8] for more details.

Most optimisation schemes use Eq. (10.15) with direction $[p^{(k)}]$ calculated using steepest descent or same quasi-Newton method. Evaluating $\alpha^{(k)}$ is the next step. At each iteration we have to decide how far in the direction of $[p^{(k)}]$ one must go. One simple way is to nominally keep $\alpha^{(k)} = 1$ and check if the following inequality is satisfied [9]

$$f([x]) - f([x^{(k)}]) \leqslant \epsilon\alpha\nabla f\cdot[p^{(k)}]$$

where ϵ is a constant chosen to be between 0 and 1. If the equality is not satisfied we reduce $\alpha^{(k)}$ till it is satisfied. A more rigorous method is to travel in direction $[p^{(k)}]$ till a minimum is reached. It is very tedious, if not impossible, to find $\alpha^{(k)}$ exactly by differentiation $\left(\text{i.e. by putting } \dfrac{\partial f}{\partial \alpha^{(k)}} = 0\right)$.

Usually $\alpha^{(k)}$, where the function is minimum in direction $[p^{(k)}]$, is found by some approximate iterative technique. In one class the range of $[x^{(k)}]$ is reduced to as narrow an interval as possible by some search technique. In the other class a quadratic, cubic or some other fit is done to f along $p^{(k)}$ and the minimum of the fit is found.

The simplest search technique is to keep dividing the range into three equal segments. First we determine that the minimum is between α_u and α_l. Any crude technique can be used to do this. Then we proceed as shown in Fig. (10.18). We evaluate f at α_l, α_1, α_2 and α_u in that order. If f increases at any α, we can say that the minimum is within the preceding two segments. This two-third part we can split into three equal parts and again find the new range. This search technique requires two extra evaluations of f at each iteration. In the golden mean search just one extra evaluation is needed at each interation. The range α_l to α_u is divided into three unequal parts, as shown in Fig. 10.19, such that

$$\frac{l}{l - x} = \frac{l-x}{x}$$

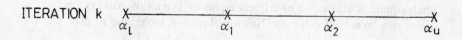

ITERATION k

ITERATION (k+1)

ITERATION (k+2)

Fig. 10.18 Search technique where range is divided into three equal parts a teach iteration. Each iteration requires two evaluations of f.

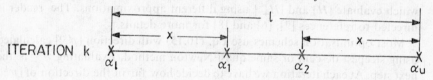

ITERATION k

ITERATION (k+1)

ITERATION (k+2)

Fig. 10.19 Golden mean search where just one evaluation of f is needed for each iteration

Solving this quadratic for $\dfrac{x}{l}$ we get $\dfrac{x}{l} = \dfrac{3 - (15)^{1/2}}{2}$

and

$$\frac{l}{l - x} = \frac{l - x}{x} = 1.618$$

Again if f increases at any α as we go from α_l to α_u, we can say that the minimum is in the preceding two segments. In order to further divide this new range into three segments just one extra point is needed. The Fibonacci search is superior to the golden mean search, but it requires that the number of iterations to find $\alpha^{(k)}$ be determined in advance.

In the quadratic fit approach we expand f around $[x^{(k)}]$ as a function of $\alpha^{(k)}$ in the direction $[p^{(k)}]$. Now f is a function of one variable. In order to avoid confusion, let us replace $\alpha^{(k)}$ by α. We can, for clarity, write

$$f([x^{(k)}] + \alpha[p^{(k)}]) = y(\alpha)$$

Let some initial guess α_0 be made for the location of the minimum of $y(\alpha)$. Expanding $y(\alpha)$, around α_0, we get

$$y(\alpha) = y(\alpha_0) + y'(\alpha - \alpha_0) + y'' \frac{(\alpha - \alpha_0)^2}{2} \qquad (10.16)$$

where $\qquad y' = \dfrac{dy}{d\alpha}$ and $y'' = \dfrac{d^2y}{d\alpha^2}$

We find α_1 from α_0 by putting $\dfrac{dy}{d\alpha} = 0$ in Eq. (10.16). This gives

$$\alpha_1 = \alpha_0 - \frac{y'}{y'_{/\alpha = \alpha_0}}$$

For the next iteration we get

$$\alpha_2 = \alpha_1 - \frac{y'}{y'_{/\alpha = \alpha_1}}$$

This is continued till convergence is reached, This quadratic fit method is the same as Newton's approach given in Eqs. (10.13) and (10.14). Equation (10.16) is similar to Eq. (10.13) except that there is only one variable.

There is a trade-off between the number of iterations to find α and the number of iterations to find the direction. If α is found very accurately, then convergence of the outer iterations to find direction would be faster.

Whatever be the optimisation technique, gradients have to be computed. (Non-gradient methods should be avoided as far as possible). Gradients can be calculated using adjoint networks. For details, the reader may see [2] and [4].

10.6 Summary

Sensitivity analysis helps in determining how sensitive a certain output, like voltage, is to small changes in the value of some circuit element. Two approaches have been described to do sensitivity analysis. The sensitivity model approach is more convenient for calculating the variation of many responses with respect to one parameter. The adjoint network approach is more convenient for finding the variation of one output with respect to many parameters in the circuit. The discussion has been restricted to DC circuits, both linear and non-linear.

Optimisation is a related topic. The circuit optimisation problem is not very different from the general optimisation problem. The topic has been merely introduced in this chapter. The problem is solved iteratively. In each iteration, one first finds the direction in which one must move to find the optimal solution. Next one must decide how far one should move in that direction. Techniques to do both were outlined in this chapter.

References

1. F.H. Branin Jr., Network Sensitivity and Noise Analysis Simplified, IEEE Trans. Circuit Theory, pp. 285-288, vol. CT-20, May, 1973.

2. R.K. Brayton and R. Spence, Sensitivity and Optimisation, Elsevier Scientific Publishing Co., 1980.
3. D.A. Calahan, Computer-aided Network Design, Chap. 5, McGraw-Hill Book Co. 1972.
4. S.W. Director and R. Rohrer, The Generalised Adjoint Network and Network Sensitivities, IEEE Trans. Circuit Theory, vol. CT-16, pp. 318-323, Aug. 1969.
5. R.K. Brayton and J. Cullum, An Algorithm for Minimizing a Differentiable Function Subject to Box Constraints and Errors, JOTA, vol. 29, No. 4 pp. 521-558, 1979.
6. S.S. Rao, Optimization: Theory and Applications, Wiley Eastern Ltd., 1978.
7. S.W. Director, Survey of Circuit-Oriented Optimisation Techniques, IEEE Trans. Circuit Theory, vol., CT-18, pp. 3-9, Jan. 1971.
8. R. Fletcher and M.J.D Powell, A Rapidly Convergent Descent Method for Minimisation, Comput. J., vol. 6, pp. 163-168, June 1963.
9. B.N. Pshenichny and Y.M. Danilin, Numerical Methods in External Problems, Chap. 2, Mir Publisher, 1978.

Problems

1. For the linear circuit given in Fig. 10.7(a) find
   (a) the column sensitivity vector

   $$\left[\frac{\partial V_A}{\partial R_4}, \quad \frac{\partial V_B}{\partial R_4}, \quad \frac{\partial V_C}{\partial R_4}\right]^t$$

   (b) the row sensitivity vector

   $$\left[\frac{\partial V_C}{\partial R_1}, \quad \frac{\partial V_C}{\partial R_2}, \quad \frac{\partial V_C}{\partial R_3}, \quad \frac{\partial V_C}{\partial R_4}\right]$$

   Use the sensitivity model approach.

2. For the non-linear circuit given in Fig. 10.9(a) find the row sensitivity vector

   $$\left[\frac{\partial V_B}{\partial R_1}, \quad \frac{\partial V_B}{\partial R_2}, \quad \frac{\partial V_B}{\partial R_3}, \quad \frac{\partial V_B}{\partial R_4}, \quad \frac{\partial V_B}{\partial k}\right]$$

   Using the sensitivity model approach.

3. Do problem 1(b) using the adjoint network.

4. Do problem 2 using the adjoint network.

5. For the non-linear circuit given in Fig. P 10.5 construct the adjoint network if V_{CE} is the output whose variation with respect to various parameters is needed

6. Show that for an independent source

   $$\frac{d(I_0, V_0)}{dE} = -I_E'; \quad \frac{d(I_0, V_0)}{dJ} = V_J'$$

   where I_E' and V_J' refer to the adjoint network and E and J are sources.

7. Show that for a linear controlled source of strength K

   $$\frac{d(V_0, I_0)}{dK} = \pm \text{ (controlling branch variable in actual network} \times$$

   (controlling branch variable in adjoint network)

8. Consider a network having n elements whose $i - v$ relationship is a function of temperature T. How would one find the sensitivity $d(V_0, I_0)/dT$ of some ouput (V_0, I_0) using the adjoint network approach ?

9. Find the minimum of the function $f(x, y) = (1 - x^2/a^2 - y^2/b^2)^{1/2}$ by
   (a) Newton-Raphson technique
   (b) Method of Steepest Descent

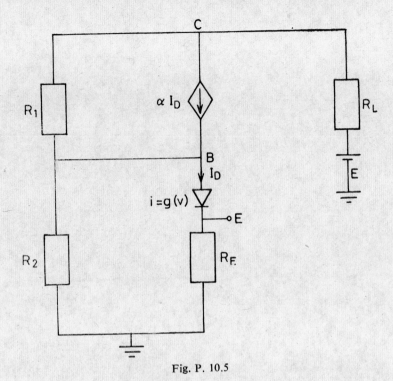

Fig. P. 10.5

In both cases start with $x = a/2$ and $y = b/2$. If $a = b = 1$ show that the method of steepest descent converges in one step.

10. Show that the ith and $(i - 1)$th directions are orthogonal in the method of steepest descent if $\alpha i-1$ is chosen exactly.

11. The DC biasing of a *CE* transistor amplifier is to be optimal with respect to the four resistors R_1, R_2, R_3, R_4. Ideally V_{CE} and I_C should be constant at the values 3 V and 1 mA respectively over the temperature range T_1 to T_2. Verify that the function to be optimised is

$$f(R_1, R_2, R_3, R_4) = \int_{T_1}^{T_2} (I_C(T) - 1 \text{ mA})^2 \omega_i \, dT + \int_{T_1}^{T_2} (V_{CE}(T) - 3V)^2 \omega_v \, dT$$

where ω_i and ω_v are weights.

Index

ERRATA

p. 35—Equation 3.2 should be as given below. Index above first summation sign should be "i" not "k"

$$a_{ij} = \sum_{k=1}^{i} a'_{ik}a_{kj}, \qquad \text{if} \qquad i < j \tag{3.2}$$

p. 41—Sentence in third line and first part of the fourth line should read "Where two elements produce the same number of fills, the element having more non-zero elements in its row/column is chosen as the pivot".

p. 47—Second to the last line should be "LEQTIF" not "LEGTIF"

p. 87—last line of the first paragraph should read "value of η is inaccurate for silicon diode"

p. 95—First line should be "R_{dc} : B—C diode non-linearity given by"

p. 97—In middle of page, in two places substitute β for B
"The ratio I_C/I_B must correspond to β_F and therefore"

"Similarly, $I_3 = \dfrac{I_S}{\beta_R}$."

p. 106—In the first equation at bottom of page, "Q_I" instead of "Q_T"

$$I_{DS} = \mu_n \frac{4Q_I}{dy} \cdot \frac{dV}{dy} = \mu_n C_{ox} \, W \quad [\ldots$$

p.107—First line below Fig. 6.15, "V_{DS}" needs to be added
"The I_{DS} versus V_{DS} characteristics can . . ."

p. 116—Eq. 6.15 should read as:

$$= \frac{2}{3} C_{ox} W l$$

$$\left[\frac{(V_G - V_D - V_T)^2 + (V_G - V_D - V_T)(V_G - V_S - V_T)^2 + (V_G - V_S - V_T)^2}{[2(V_C - V_T) - V_D - V_S]} \right]$$

p. 128—Fourth line of second paragraph, "diode" instead of "transistor"
". . . The Schottky diode introduced in Fig. 7.3(b) prevents"

p. 135—Second last line of first paragraph, "," should be deleted
"Poon parameters for the transistors in a commercial IC are not easy to"

p. 192—Eighth line of second paragraph, should read
". . . We make an initial guess for all m variables"

p. 194—Third line of second paragraph should read as
"G—J takes m iterations . . ."

p. 197—First line of equation 9.7, "C_A" instead of "C_{AA}"

p. 201—Sixth line of third paragraph, "Gauss" to be deleted

p. 203—Section 9.5 should read as
"9.5 Waveform Relaxation [11]"

p. 210—Fourth line of second paragraph, should read as
". . . They use iterative techniques like Seidel-Newton to"

p. 219—Second last line, should read as
"For AC linear circuits capacitors and inductors may vary in value . . ."

p. 229—Third line should be
"technique of analysis would have to be used. For example, modified node"

p. 229—Last line, should read as
"modified node admittance matrix of the linearised . . . "

p. 236—Equation 10.13, "∇" instead of "∇" at end

p. 238—Figure caption of Fig. 10.18, read
". . . three equal parts at each"